ROUTLEDGE LIBRAR
SOCIAL AND CULTURA

Volume 3

THE GEOGRAPHY OF CRIME

THE GEOGRAPHY OF CRIME

Edited by
DAVID J. EVANS
AND DAVID T. HERBERT

LONDON AND NEW YORK

First published in 1989

This edition first published in 2014
by Routledge
2 Park Square, Milton Park, Abingdon, Oxfordshire OX14 4RN

and by Routledge
711 Third Avenue, New York, NY 10017

First issued in paperback 2015

Routledge is an imprint of the Taylor & Francis Group, an informa business

British Library Cataloguing in Publication Data
A catalogue record for this book is available from the British Library

ISBN: 978-0-415-83447-6 (Set)
ISBN: 978-1-138-98936-8 (pbk)
ISBN: 978-0-415-73154-6 (hbk) (Volume 3)

The Geography of Crime

Edited by David J. Evans
and David T. Herbert

ROUTLEDGE
London and New York

First published 1989
by Routledge
11 New Fetter Lane, London EC4P 4EE
29 West 35th Street, New York NY 10001

Printed and bound in Great Britain by Mackays of Chatham PLC, Kent

British Library Cataloguing in Publication Data

The Geography of crime.
1. Crime. Geographical factors
I. Evans, David J. (David John), *1947-*
II. Herbert, David T. *1935 Dec 24-*

364.2′2

ISBN 0-415-00453-5

Library of Congress Cataloging-in-Publication Data

The Geography of crime / edited by David J. Evans and David Herbert.
p. cm.
Bibliography: p.
Includes index.
ISBN 0-415-00453-5
1. Crime and criminals. 2. Man — Influence of environment.
3. Environmental psychology. 4. Crime prevention. I. Evans, David J., 1947- . II. Herbert, David T.
HV6150.G39 1988
364.2′2–dc19 88-23634
CIP

CONTENTS

PLATES

FIGURES

TABLES

CONTRIBUTORS

Slak Bartnicki, Faculty of Geography and Regional Studies, University of Warsaw, Poland

Trevor Bennett, Institute of Criminology, University of Cambridge

Pat Brantingham, Department of Criminology, Simon Fraser University, British Columbia

Alice Coleman, Department of Geography, King's College, London

Norman Davidson, Department of Geography, University of Hull

David Evans, Department of Geography and Recreation Studies, North Staffordshire Polytechnic

Nick Fyfe, Department of Geography, University of Cambridge

Keith Harries, Department of Geography, University of Maryland, Balitmore County, Baltimore, Maryland, USA

Linda Harvey, Department of Social Administration, University of Manchester

Kevin Heal, Home Office Research and Planning Unit

David Herbert, Department of Geography, University College of Swansea

Mike Hough, Home Office Research and Planning Unit

Gloria Laycock, Home Office Research and Planning Unit

Helen Lewis, Home Office Research and Planning Unit

John Lowman, Department of Criminology, Simon Fraser University, British Columbia

Rob Mawby, Department of Social and Political Studies, Plymouth Polytechnic

Ken Pease, Department of Social Administration, University of Manchester

George Rengert, Department of Criminal Justice, Temple University, Philadelphia
Stephen Stadler, Department of Geography, Oklahoma State University
Susan Smith, Centre for Housing Research, University of Glasgow

PREFACE

This collection of essays on the 'geography of crime' has its origins in two conferences held in 1986. The first was held at North Staffordshire Polytechnic with the support of that institution and the second formed part of the programme of meetings organized by the Institute of British Geographers' Study Group in Urban Geography. Most of the contributions to these conferences have been included in this volume. With the preparation of a new text, however, the opportunity was present to draw in a wider range of contributors and to aim for a book which more fully represented the burgeoning interest in the 'geography of crime' and something of the international spread of that interest. The initial focus on the spatial ecology of crime has been replaced by a more diverse set of approaches which are concerned more explicitly, for example, with spatial behaviour, with images and fears of crime, with the workings of the criminal justice system and with the relevance of research for crime prevention policies.

This text offers a balanced and up-to-date set of perspectives on the 'geography of crime'. Its contributors are drawn from several parts of the world and include both established figures and others who are only now beginning to make an impact on the research literature. It offers both students of crime and delinquency and practitioners of criminal justice and crime prevention an overview of this important field of research and insights into the future directions which it is likely to follow.

David J. Evans
David T. Herbert

Chapter One

CRIME AND PLACE: AN INTRODUCTION

David T. Herbert

This is in many ways a timely book. The geography of crime has come of age in the 1980s and both the quickening of pace in terms of research and the willingness to move into new kinds of topical areas reflect this. There is now a range of content it would not have been possible to depict a decade ago and also a blurring of the boundaries between kinds of analyses conducted by geographers and those which are typical of criminology as a whole. A 'geography' of crime will always carry its particular hallmarks of an interest in spatial structures, in environmental associations, and in the special qualities of place, but any tendency which might have existed for spatial chauvinism is now far less evident than it might have been in the past.

Geographers studying crime have always recognized some early roots in the kinds of cartographic criminology which appeared during the nineteenth century as magistrates, government statisticians, and others demonstrated the fact that patterns of crime were unevenly spread across cities and regions. Again, the social ecology of the Chicago school which was so influential in the inter-war period contained basic statements on the occurrence of crime or delinquency areas within the city which have had lasting methodological effects. A modern geography of crime inherits these traditions in a critical way but also moves towards a wider range of concerns and to quite different methodological and conceptual positions. The purpose of this introduction is to sketch the broad areas of current concern and to identify the kind of research agenda which faces geographers concerned with the study of crime. These objectives will be achieved both by some initial comments on the relationship between crime and place and also by introducing in a systematic way the contributions which make up this text.

Traditional concerns

The term 'traditional' is taken to include the twin, interrelated traditions of cartographic criminology and spatial ecology. The earlier 'geographies' of crime were cast in this tradition with a focus on regional variations in different types of crime and of justice. Other studies focused on crime in the city and explored ways of defining crime or delinquency areas and relating these to a range of social indicators of the urban condition. The theories acknowledged and to a lesser extent used in these early studies were those of the social sciences at large. Social disorganization and delinquent sub-culture, for example, were attractive because they could be stated in 'area' form, that is they were typical of some kinds of urban neighbourhoods. Interestingly, when The Urban Criminal was published as one of the major empirical studies of the 1970s (Baldwin and Bottoms 1976) it turned towards these theories, especially social disorganization, as a source of understanding. This serves to indicate that traditional concerns are by no means unworthy of continuing research and development. Again, The Urban Criminal was written by criminologists not geographers and has offered a useful bridge to the wider field of study.

The mapping and recording of crime is of continuing value to geographers; as the basic tools and methods of cartographic representation improve so will geographers of crime be better placed to use these in their own research. A first step towards employing these new technologies and, for example, using computer graphics in relation to crime statistics will be a more systematic way of coding such data to a geographic or area reference system. Pyle (1974) proposed a way of forming data recording systems for crime such that they would allow rapid and easy representation in spatial form; the technology and the awareness clearly exist; all that is needed is the implementation. As the computerization of police records progresses it is imperative that it tackles the issue of spatial coding and of proper maintenance of data to allow pattern analysis over time and space.

Crime and place

The concept of place has special meaning in geography which dates most explicitly from the work of French geographers such as Vidal de la Blache (1913) who argued that geography was the science of places rather than of men. Place became somewhat submerged in the 1960s with the focus on locational analysis which was more concerned with the geometry of space than with the meanings of place. The emergence of humanistic geography promises a new era with its belief that place has to be much more than position in space, rather a territory to which people attach values. As Relph (1976:141) argues:

> Places in these terms are fusions of human and natural orders and are significant centres of our immediate experience of the world. They are defined less by unique locations than by the focusing of experiences and intentions on to particular settings.

Places can thus be seen as subjective and qualitative concepts, underpinned by ideas such as sense of belonging, shared values and images, and common concerns. There are many ways to view and to use the concept of place but geographers concerned with crime should perhaps give some increasing emphasis to place in terms of this kind. As geographers identify clusters of crime and regard them as 'delinquency areas', the task is then to understand the ways in which such 'places' have emerged and the characters which they have assumed. Social ecology still provides some kind of framework for this type of approach and the key here is to draw more heavily on that part of its inheritance which relates to its rich 'humanistic', ethnographic tradition rather than to the more mechanical aspects of ecological association. There are examples in the literature of the geography of crime which show how ideas of place can be used in these ways.

Herbert (1976) provides one such example in relation to the persistence of delinquency areas in Cardiff from which young offenders seemed to be drawn in disproportionate numbers. The research strategy allowed residential areas to be classified according to a series of objective indicators and then to identify situations in which objectively similar places were different in terms of the incidence of offenders. Evidence here pointed to the fact that there were differences in values, attitudes, and related forms of behaviour which closely mirrored the levels of delinquency. The places were different subjective environments and these differences held significance for the existence of crime. Places may also differ in terms of levels of vulnerability to crime, some seem to attract offenders much more than others. Herbert and Hyde (1985) identified a number of areas which suffered from residential burglary in Swansea and showed that place differences could underlie such vulnerability. Testing a number of area hypotheses, they were able to show that rates of burglary could be understood in these terms.

Another thrust towards giving place a more central position in the study of crime was provided by Brantingham and Brantingham (1981) in their exposition on 'environmental criminology'. They argued that any crime consists of the four dimensions of the law, the offender, the target, and the place. All of these have essentially to be integrated into any overall understanding of crime but each is capable of focused study in its own right. Environmental criminology is the study of place, the fourth dimension, which is defined as the discrete location in time and space at which the other three dimensions intersect and a criminal event occurs. The scale of place is that of neighbourhood or local community within the city; the focal interest of environmental criminology is the offence rather than the offender and the characteristics of place are central to an understanding of why a specific event occurs. Environmental criminology provides a useful framework for geographers and a link with the wider field of study. A further, closely connected focus on place can be found in the development of situational crime research with which crime-prevention policies have become increasingly associated. When criminologists talk about the settings or situations in which crimes occur and the opportunities which some environments offer to offenders, they are essentially seeking to define places in ways familiar to geographers. Attempts to codify those features of situations in which criminal events occur are in effect attempts to measure aspects of place; crime-prevention policies which aim at situational change to reduce vulnerability are in effect trying to modify place. This type of approach has a direct line to defensible space ideas and their beliefs that crime can be tackled at a local level.

This place theme can also be related to some of the

more recent topical areas within the geography of crime. Over a decade ago now, David Ley (1974) argued that stress areas could be identified in parts of the inner city where normal patterns of behaviour could be adversely affected and constrained by the fact that people felt unsafe. During the 1980s an awareness of the significance of fear of crime has gained force and researchers have begun to examine the concept of place in this context. Fear undoubtedly affects some groups and individuals more than others but it is also evident that there are place problems as well as people problems. If there are areas emerging in our cities where people are afraid to go out at night or where residents feel they cannot leave their property unprotected then these are places to be studied and understood.

Other trends and a research agenda

The place focus offers one strong theme for the future development of a geography of crime but there are others. Spatial behaviour and behaviour in space offer themes well understood in some areas of geographical research but not at all well explored in the context of crime. The theme of journey to crime has been studied and a number of generalizations are available on typical distances travelled by different types of offenders, etc. Behavioural characteristics of known offenders have been studied but this research has concentrated on the 'professional' type of offender being held at some place of custody; much less is known about the opportunist offender towards whom many of the present crime-prevention initiatives are being aimed. Some of the early work on offenders' images of the city needs replication in behavioural research. Again, a focus in behavioural studies on the offenders should be balanced by a greater emphasis on the victims and potential victims whose behaviour in space can be highly significant. Behavioural studies have some way to go in the geography of crime. There is much to be done in relation to offenders and offences, new initiatives to be taken in the study of victims and a greater understanding to be reached on the ways in which fear of crime affects people's lives.

Geographers have begun to show much more explicit interest in the criminal justice system as a research theme. One of the qualities of this collection of essays is that it does offer several strong examples of ways in which this shift of interest is being accomplished. The markers have been down for some time. Geographers have noted the variability within the criminal justice system and the ways in which sentencing patterns, for example, vary considerably among courts in different parts of the country. There is also an awareness of the definitional problems associated with crime statistics affecting both their production and the ways in which they are used. The shifts within criminology from positive science to the sociology of deviance and the radical criminology of the 1970s have been well noted. Of more direct interest to the crime and place theme was the kind of message contained in Shumsky and Stringer's study (1981) of vice areas in San Francisco at the turn of the century. Their basic contention was that the places at which vice was practised were fashioned more by the political wills of the city fathers and the ways in which these were interpreted by the police rather than by the 'actors' themselves. Places associated with vice emerged and disappeared, concentrated or dispersed in response to the workings of the criminal justice system. This interest in the impact of the criminal justice system upon crime has been extended in other ways and with other questions. How do policing policies affect crime within the city? If policing intensifies do crime rates fall or do they rise because recording of crime intrudes into the 'dark area'? How do the police and the courts use their discretionary powers and how are these reflected in geographical space? Another issue of modern times is that of police accountability. This raises questions of the links between police and local democracy, between police and levels of government. These police roles are subject to legislative changes which have implications for urban communities.

This brief introduction identifies some of the key elements which have begun to typify a geography of crime in the 1980s and which are forming the ingredients for a future research agenda. The geography of crime is widening its conceptual bases to encompass selected aspects of both structuralism and humanism. The former is revealed by the increased interest in the criminal justice system and the latter by the concern with place. In these ways a 'geography' of crime begins to resemble more closely the breadth of topical concerns which typify the wider field of criminology. In the chapters which follow some of the elements of this 'sea-change' will be evident. Our collection contains elements of the old, examples of the new, and some of the portents for future research in the geography of crime.

Organization and content of this text

Data have always posed particular problems for research in criminology and the issue recurs at many points in this text as various contributors feel the need to qualify the kinds of insights they can offer into an aspect of the 'crime problem'. We begin the book therefore with a report upon a relatively recent phenomenon in Britain which will have an increasing significance for the study of crime and the use of criminal statistics. The phenomenon is the national crime survey, available in the United States since 1972 and being used in Britain during the 1980s. As Mike Hough and Helen Lewis show in their chapter, there are several other countries such as Canada, Australia, the Netherlands, and Eire, which now have crime surveys and the acceptability of the approach in Britain is shown by the fact that there will be three national crime surveys in the 1980s.

These surveys are not perfect by any means but do represent major advances in the gathering of data which gives a far more complete and insightful picture of the incidence of crime and its effects than is afforded by the official criminal statistics. We already, for example, have some more definitive information on reporting rates which range from 9 per cent of thefts (valued less than £26) from cars to 99 per cent for thefts of cars; over all types of crime the reporting rate stands at 38 per cent. As the databank is enhanced by successive national crime surveys, far more reliable estimates of the incidence of different types of crime will be possible together with a better understanding of the so-called 'dark area' of offences which occur but never become part of the official record. The British Crime Surveys are being used not merely to enhance basic data but also to ask and answer other types of questions such as the levels of risk and of fear of crime amongst different segments of the population.

These official national crime surveys have been augmented by a number of major local surveys and one of the best known of these is the Islington Crime Survey (Jones, McLean, and Young 1986). Conducted in the London Borough of Islington, which is one of the worst affected areas in Britain, this survey shows that over a twelve-month period a staggering 1 in 2 households could expect to be the victims of some kind of crime and about half of those who experience crime would do so more than once. About 1 in 11 households would be burgled and 1 in 10 of these burglaries would be cleared up with very little chance of goods being recovered. It is not surprising that insurance is expensive and difficult to get for people living in some parts of Islington. Figures on assault were equally dramatic and especially for assaults on women. The overall figure was that 1 in 40 women would be victims of sexual assault over a twelve-month period but that this would rise to 1 in 11 for young, white women and 1 in 6 if they frequently went out in the evenings. Lastly and of interest in terms of some of the contributions which follow in this book, there were significant insights on levels of fear of crime. Over half the women avoided going out after dark and 25 per cent never went out alone at any time. Surveys of this kind add significantly to our understanding of the extent of the crime problem and the impact it has on people's lives; the national crime surveys, augmented by major local surveys, promise to provide far better bases for both academic research and action.

Many geographical studies of crime in the 1980s have continued to focus on the environments in which crimes occur and, to a lesser extent, on the environments from which offenders come. The term 'environment' has become more various in its interpretations and it is now more common to distinguish between built and social environments, between objective and subjective, and between resource and physical environments. These differing kinds of environments and the relationship to crime are reflected in several of the contributions. Keith Harries and Stephen Stadler return to one of the oldest themes in human geography, the relationship between climatic variations and human behaviour. They study a form of crime for which American cities have attained some notoriety - violence and criminal assault - and examine its relationship with heat stress in Dallas. Over an eight-month period, some 4,309 aggravated assaults were reported to the Dallas police and analysis showed a positive association between the incidence rate and an index of discomfort based on local climatic conditions. There was an inverse relationship between assault and socio-economic status but the hypothesis that a threshold of discomfort could be reached beyond which all activities, including crime, ceased, could not be confirmed. This evidence leads to the conclusion that climatic conditions may well exacerbate stress and be crime-related; a new approach which focuses on general measures designed to improve public health and life quality may yet have positive returns in terms of the control of acts of violence. Norman Davidson continues the theme of violence in its environmental settings but widens the terms of reference to the settings or situations in which acts of violence occur. These settings have a range of distinguishable properties and do contain clear variations.

There are 'place' variations with differences between types of residential areas, time variations, and 'people' variations with specific types of people at greater risk. A typology of situational criteria is becoming attainable which should help in the understanding of crimes of violence. David Evans changes the focus to residential burglary and the different kinds of conditions in which it is most likely to occur. The risk factors can be isolated and categorized or, alternatively, the types of residential areas likely to be more vulnerable can be identified.

The built environment as an influence on criminal behaviour was thrown into sharp focus by Oscar Newman (1972) and his thesis of defensible space. This thesis proved very attractive to those faced with the realities of coping with crime, not least because it offered a number of practical and applicable proposals on ways to design out some forms of crime. The thesis survives despite the academic criticisms which point towards its conceptual shortcomings. Alice Coleman follows some of the basic Newman ideas through in her detailed researches concerned with parts of inner London and its housing estates. By combining a set of key indices, Coleman produces an Index of Disadvantage which forms a measure of the extent to which different parts of the city suffer from defects in design and quality of the environment. The basic contention is that crime rates will be higher in those places which suffer the most from disadvantagement and, further, that if measures are implemented to reduce the index scores, then crime rates will fall and a better quality of life will ensue. Coleman acknowledges some of the criticisms of the 'improvement through design' approach but argues that the theory deserves to be subjected to rigorous testing through its incorporation into policy initiatives in some of the areas most affected by crime and vandalism.

This type of analysis of the impact of the built environment upon crime rates is also evident in Slavak

Bartnicki's study of Poland. This is a rather more broadly based contribution which has value in outlining temporal trends and regional variations in the incidence of crimes in an East European country about which little has previously been written in this context. Bartnicki shows that within Warsaw there are sharp neighbourhood differences in crime rates with some 'problem areas' of the kind familiar to western criminologists. There are also neighbourhood variations in the levels of fear of crime, and the relationships these hold to actual crime rates and awareness of crime are scrutinized. Inhabitants of Warsaw respond to fear of crime by adopting a range of situational crime-prevention measures, notably bars or grills across doors and windows. Bartnicki shows how the adoption of these measures is much more typical of some types of urban settings than of others. Although fears of being victimized are widely held and influence behaviour in space, they are out of proportion to the real levels of risk.

Behavioural studies have assumed considerable significance in human geography at large and George Rengert argues that they have major roles to play in the study of crime and criminal behaviour. The spatial decision-making of individuals can be analysed both as spatial behaviour and as behaviour in space and the latter, in particular, has many applications in relation to the analysis of offenders and their victims. Among the more neglected themes which geographers of crime could usefully incorporate into their researches is the concept of awareness space. Women, for example, have more limited awareness spaces than men, inner-city youths than suburban youths. Similarly the behavioural concepts of search space and revealed space are well worth application in crime studies. Any focus on opportunities must involve close attention to different types of barriers and constraints upon criminal behaviour and the constraining effect of correctional practice offers an especially promising field of enquiry. The approach of interviewing known offenders is well established in criminological research and several recent studies have added to our understanding of criminal behaviour. There are hazards in the method: accuracy of information given, for example, and also the extent to which any sample of known offenders, particularly if drawn from groups held in prison, is representative of the offender group as a whole.

Trevor Bennett reports on his own study of burglars'

perceptions and decision-making. His sample consists of imprisoned burglars with long records of offences and he draws his information from both their reaction to videos of residential settings and semi-structured interviews. His objective with both these methods is to identify which situations burglars seek and which they avoid. From the videos the burglars seemed to be attracted by cover of any kind, easy forms of access, and any signs of affluence; from the interviews they were put off by signs of occupancy, visible neighbours, possibilities of being overlooked, and any good locks or other security devices. This kind of approach will add to our understanding of offender behaviour, the task now may be to achieve better awareness of the decision-making of more opportunistic offenders. If it is important to understand the behaviour of offenders, it is also important to be aware of the attitudes of the actual and potential victims. Fear of crime is a state of anxiety which varies spatially but by no means in close accord with the actual incidence of crime.

Susan Smith looks for those environmental cues which may predispose a person to feel fear of crime and nominates the media, the law enforcement agencies, the incivilities of the built environment, and the deterioration of family life. Age, gender, household structure, and tenure are more individual factors which affect levels of fear and these relate to feelings of powerlessness and lack of resources. Smith argues that any policy response should be sensitive to the needs of specific types of people as well as to the aggregate milieu.

As a 'geography of crime' develops so the need to examine its conceptual bases acquires increasing significance. A 'geography' of crime may belong to a particular subset of criminology but our efforts should be towards contributing to the field of study as a whole rather than circumscribing the subset in any rigid way. John Lowman directs this argument to the concept of social control. His belief is not in the primacy of the social-control perspective over all others but rather that it must form part of any study of crime including those centred on spatial structures. Social control involves a concern with the action of the state and other agencies in mobilizing resources against anticipated deviance. The police clearly act as one agent of social control and the roles of the police have recently come into sharper focus with the tensions between their autonomy as 'enforcers' of the law and their responsibilities towards local democracy and to local and central government.

Rob Mawby takes a longer view of the role of the police and reviews a number of stages at which the relationship of police to academic studies of crime has taken different forms. Initially the police were peripheral to the understanding of crime and this was followed by a period of reliance on police statistics and subsequently more questioning of both the statistics per se and the ways in which the police and others were involved in the definition of crime and their use of discretion in the recording of offences. Scrutiny of styles of policing now forms an established research interest and various policing strategies of crime control are examined, especially as they relate to local problems. Mawby has adopted a position in the debate on the extent to which police practice affects crime rates.

The relative autonomy of the police and their links with local democracy are the central concerns of Nicholas Fyfe's chapter. The idea of establishing local community/police consultative committees across England and Wales can be found in the Scarman Report which followed the riots of 1981 and this is now being implemented. Police are required to 'listen' to the local community on a regular basis; the issues are of defining the local community and its representation for these purposes. Although consultative committees are now appearing, it has not been an easy process. Local political attitudes vary and cannot be categorized simply on party lines. There are tensions which arise from these different attitudes and revolve around wider issues of central government, local government, local pressure groups, and the police. The virtual eclipse of the police authority, which had constituted the traditional link with local democracy, has exacerbated the issue of who meets who in the consultative process. The incorporation of laws, justice, and sentencing procedures into the purview of the spatial analysis of crime is not new but there is potential for more research in this area.

Linda Harvey and Ken Pease offer a range of insights into the ways punishment sanctions vary across the agencies responsible. They remind us that the data deficiencies which hinder most forms of research in criminology are present here and crime known to or cleared up by the police does not provide an adequate backcloth against which to assess differences in punishment. Alongside the courts, the police exercise certain discretionary powers and cautioning is no longer numerically a trivial justice procedure. Cautioning practice varies with a host of intervening variables such as age, gender, and criminal record. Inconsistencies among courts defy single-stranded explanations. Evidence points towards local effects and to the significance of key gatekeepers in the system such as the court clerks. Harvey and Pease move towards a conclusion that the sources of criminal justice differences are primarily the perceptions and preferences of the agents of the local justice system; the more fully aware those agents are of local fears and perceptions the better.

The crime-prevention model is an attractive one for a society concerned with the rising 'cost' of crime for reasons similar to those which make a public health model attractive to a society concerned with the rising costs of health care. A preventive policy aims to reduce the risks of crime, to form safe living environments, and to create situations in which potential criminals are least likely to succeed. Gloria Laycock and Kevin Heal review recent British experience of crime prevention and see the 1960s as the time when central government began to change the balance towards tangible support for crime prevention. Crime prevention remained the poor relation in the police service through to the 1980s with less than 0.3 per cent of the police being designated as crime-prevention officers. During the 1980s crime-prevention initiatives have started at national levels and there is an intention to see crime as a problem for the community as a whole. There are specific measures such as new roles for crime-prevention panels, design features for car security, and schemes such as Neighbourhood Watch which recognize that the public experience, observe, and report crime, and that that role should be maximized.

The notion of heightening public awareness of crime is a double-edged sword: to make more aware is to cause more concern and there is a fine balance to be struck in these kinds of crime-prevention policies. Pat Brantingham reminds us that crime prevention has long been the main goal of the North American justice system. There are many well-known and tested programmes of crime prevention and applied research is only more recently focusing on the evaluation process. The quest for typologies has produced many kinds of answers. Various types of interventionist strategies are possible and centre on the criminal event and its stages of decision, search, and actual commission. These interventions may take various forms from target hardening to community watch to policing practice. Standard 'off-the-shelf' packages for crime prevention remain basic answers but research will refine the requirements in more specific terms of situation, time, and place.

In conclusion

The contributors to this book cover a wide range of insights into the geography of crime. They rehearse established approaches and offer some which are new. They offer new thoughts on the conceptual bases from which geographers work and also identify and exemplify new topical areas in which the discipline can make a valuable contribution. We are not concerned to isolate an area of study and say this is a 'geography' of crime; we are concerned to show that there are missing dimensions to many aspects of criminological research and that geographers, with their interests in space and place, can enrich the wider study by emphasizing these dimensions.

References

Baldwin, J. and Bottoms, A.E. (1976) The Urban Criminal, London: Tavistock

Brantingham, P.J. and Brantingham, P.L. (1981) Environmental Criminology, London: Sage

de la Blache, Vidal, (1913) 'Des caracteres distinctifs de la geographie', Annales de Geographie 22, 124, 289-99

Herbert, D.T. (1976) 'The study of delinquency areas: a social geographical approach', Transactions, Institute of British Geographers 1, 472-92

Herbert, D.T. and Hyde, S.W. (1985) 'Environmental criminology: testing some area hypotheses', Transactions, Institute of British Geographers 10, 259-74

Jones, T., Maclean, B. and Young, J. (1986) The Islington Survey, Aldershot: Gower

Ley, D. (1974) The Black Inner City as Frontier Outpost, Association of American Geographers Monograph, Washington

Newman, O. (1972) Defensible Space, New York: Macmillan

Pyle, G.F. (ed.) (1974) The Spatial Dynamics of Crime,

University of Chicago, Geography Research Paper 159
Relph, E. (1976) Place and Placelessness, London: Pion
Shumsky, N.L. and Springer, L.M. (1981) 'San Francisco's zone of prostitution, 1880 to 1934', Journal of Historical Geography 7, 71-89

Chapter Two

COUNTING CRIME AND ANALYSING RISKS: THE BRITISH CRIME SURVEY

Mike Hough and Helen Lewis

The British Crime Survey (BCS) is a large sample survey whose main purpose is to provide an index of crime. This paper falls into two parts. Firstly it discusses the rationale of crime surveys and the design of the BCS and presents some of its main findings about the extent of crime. It goes on to present a conceptual framework for analysing the distribution of crime, illustrating this with findings on the risks of burglary.

The rationale for crime surveys

Criminologists have always known that there is a large 'dark figure' of crimes which never find their way into police records. They have also been aware, therefore, that statistics of recorded crime can only be used as a measure of the extent of crime if the proportions of crimes reported to and recorded by the police remain constant over place and time. However, despite the improbability of a constant 'dark figure', recorded crime statistics have for generations been pressed into service not only to measure the actual workload imposed on the criminal justice system, but also to show the extent of crime. In the absence of alternatives, it is not really surprising that these statistics have been used in this way. What is needed ideally, of course, is some measure of crime collected independently of processes designed to control crime. It is only recently - with improved facilities for computer analysis and better survey methodology - that crime surveys have made such measures possible. The actual motive force - the political spur - for the introduction of the British Crime Survey was the belief that statistics of crimes recorded by the police often misled people about the rate of increase in crime.

Surveys of victimization - crime surveys for short - typically question a randomly selected sample of the population about their experiences as victims of various crimes. When grossed up, the results yield estimates of the extent of these crimes for a particular year, while surveys repeated at regular intervals can help show crime trends. National crime surveys have now been carried out in several countries including the United States (annually since 1972), Canada, Australia, the Netherlands, and the Republic of Ireland. In this country, the General Household Survey (GHS) has carried questions on residential burglary intermittently since 1972. The British Crime Survey (BCS) was first carried out in 1982 in England and Wales (Hough and Mayhew 1983) and in Scotland (Chambers and Tombs 1984); the second sweep was conducted two years later - but not in Scotland - (Hough and Mayhew 1985); and a third in 1988. A number of local surveys have also been carried out in Britain, notably in Nottinghamshire and adjacent counties (Farrington and Dowds 1985), in Merseyside (Kinsey, 1984) and in Islington (Jones et al. 1986).

The great strength of crime surveys is that their image of crime has not been refracted through the workings of the criminal justice system. There are many counter-balancing limitations see for example, Skogan 1981; Sparks 1982. They can only uncover crimes which have clearly identifiable people as victims - they cannot easily count crimes against organizations (such as company fraud, shoplifting or fare evasion) or 'victimless crimes' such as drug abuse or some sexual offences. It is also well established that people fail to report to interviewers all the relevant incidents which they have experienced within the 'recall' period, and that they also report incidents which had in fact occurred earlier. In addition, there is some evidence of systematic 'response bias': for example, better-educated respondents seem more adept at recalling relevant events at interview and middle-class respondents seem more ready than others to define certain classes of incident as assaults. It must also be remembered that crime survey findings are only based on a sample of the population and error may arise because of this: the percentage sampling error is particularly large for relatively rare crimes such as robbery and rape. Since the extent of the error depends on sample size, there has to be a trade-off between the expense of a survey and its precision. Crime surveys always seem to their sponsors to cost too much (each sweep of the BCS cost in the region of £240,000) while the researchers always feel that the samples they are permitted are too small.

Finally, there are some important conceptual issues to do with the nature of incidents counted by crime surveys - are these crimes as defined by criminal law? Or according to the definitions embedded in police practice? Or as popularly defined - however that might be? Crime surveys employ more inclusive definitions of the offences with which they are concerned than do the police or - quite possibly - the public. To summarize a complex argument, statistics of recorded crime count incidents reported to the police which could be punished by a court and <u>should</u> occupy the attention of the criminal justice system. Both public and police inevitably make value judgements in deciding whether or not an incident requires the attention of the criminal justice system and the resultant count of recorded crime reflects these judgements. The BCS, on the other hand, counts a category of actions which <u>could</u> be punished, according to the letter of the law, regardless of the value in doing so. Crime surveys, therefore, count a broader and more value-free category of incidents than police statistics - of necessity if they are to be able to detect shifts in reporting and recording practice over place or time.

Many crime surveys including the BCS do not simply count crimes but also collect more detailed information about both the offences and the victims. Police statistics to date provide surprisingly little information about these areas. By collecting this additional information about crimes and victims crime surveys can offer - at least for the offences they cover - a fuller picture of which people are at risk, why this may be so, as well as an idea of the nature of both reported and unreported offences. Although the original purpose of conducting crime surveys was the fairly narrow one of crime measurement they have paid their way quite considerably with information of this sort on risks of victimization. The BCS has also collected information on a wide range of other crime-related topics including fear of crime, experience of the police, and attitudes to punishment.

The design of the British Crime Survey

The British Crime Survey was developed by the Home Office Research and Planning Unit in collaboration with the survey companies who carried out fieldwork - Social and Community Planning Research for the first sweep and NOP Market Research for the second (Wood 1984; NOP 1985). In both sweeps one person aged 16 or over was interviewed in 11,000 households in England and Wales: in the first BCS a further 5,000 were interviewed in Scotland. Respondents were selected using the Electoral Register as a sampling frame and response rates of 80 per cent and 77 per cent were achieved in the two surveys.

All respondents in the two surveys answered a main questionnaire which 'screened' people to see if they or other members of their household had been victims of offences. For personal crimes - assaults, robberies, thefts from the person, other personal thefts, and sexual offences - respondents were asked only about their own experience. For household crimes - burglary, thefts of and from vehicles, vandalism, and theft from the home - they were also asked about the experience of others in their household. Details of each incident revealed by the screening questions were collected on victim forms (up to a limit of four forms). All victims and two of out five non-victims also completed a follow-up questionnaire. In the first sweep this covered fear of crime, contact with and attitudes to the police, routine activities and lifestyle, and self-reported offending; in the second sweep, it covered fear of crime and attitudes to punishment, crime seriousness, crime-prevention measures, and self-reported offending. Demographic details were collected from all respondents.

The 'dark figure' of unrecorded crime

By combining information from the BCS with police statistics, estimates can be derived of the proportion of crimes reported to and recorded by the police. Based on estimates from the 1984 BCS of offence rates for offences with counterparts in police statistics, Figure 2.1 shows the numbers of unreported incidents, those which were reported but not recorded, and those which found their way into police records.

A large proportion of incidents went unreported for all

Figure 2.1 Unreported, reported but not recorded, and recorded incidents from the 1984 British Crime Survey

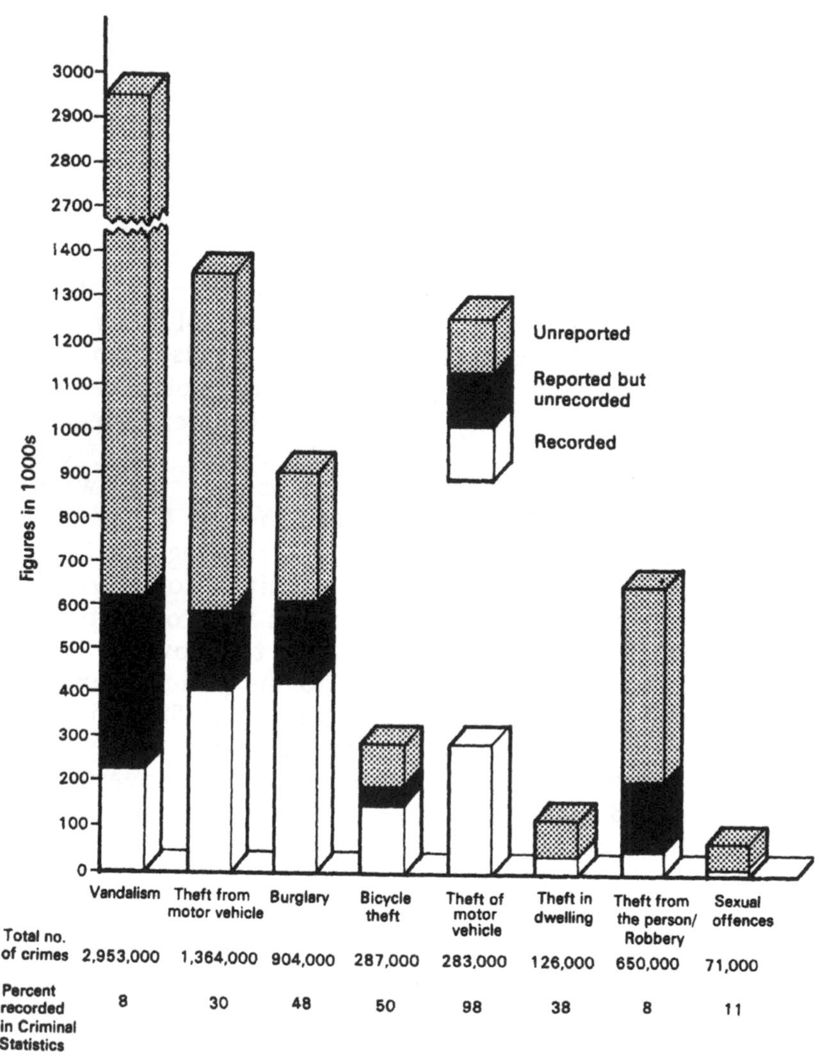

categories of crime except vehicle theft. The overall reporting rate for crimes shown in Figure 2.1 was 38 per cent. (Rates for selected categories are shown in Table 2.1).

Victims who said they had notified the police were asked why they had done so: responses were balanced between self-interest (recovering goods, making insurance claims etc.) and a sense of obligation ('You ought to'. 'It was a serious crime' etc.). Most 'non-reporters' gave as their reasons either the triviality of the incident or the fact that the police would be able to do little about it - however there was also a significant number of serious incidents which went unreported. In general reporting to the police turned - almost by definition - on perceptions of offence seriousness, but overlying the victims' sense of what properly was or was not grist to the mill of formal justice were also calculations of the costs and benefits to themselves in reporting, and of the chances that the police could actually achieve anything if informed.

As part of the follow-up questionnaire respondents were asked how serious they felt various offences were. As Table 2.1 shows overall crimes which were well reported were those with high 'seriousness ratings' while those with lower reporting rates were generally those which were rated less serious. There were some anomalies, however. Bicycle thefts and thefts from cars were better reported than their average seriousness scores would suggest - presumably because victims wanted to recover property or file insurance claims. In contrast, sexual offences had very low reporting rates but the highest ratings for seriousness. (This is consistent with the idea that women are reluctant to bring in the police over sexual offences, although it might reflect a quirk of sampling since so few sexual offences were uncovered by the survey.)

Figure 2.1 shows that many of the offences reported to the police do not get recorded as crimes - or do not get recorded in the crime categories shown. Overall, the police would appear to record two-thirds of the property crimes known to them. However, the BCS can only offer imprecise estimates of this 'recording shortfall' both because of errors in the survey estimates and because of the difficulties in comparing like with like when matching BCS offence categories to those used by police statistics. Nevertheless, the first and second surveys yielded similar estimates and there can be little doubt that many incidents reported to the police are not recorded in those crime categories suggested

Table 2.1 Percentage of crimes reported to police and seriousness ratings for selected crime categories

	% Reported	Average rating for seriousness
Well-reported crimes		
Theft of cars	99	7.6
Burglary with loss	87	7.9
Bicycle theft	68	4.7
Vandalism over £20 to the home	62	7.2
Wounding	60	7.7
Robbery	57	7.6
Less well-reported crimes		
Attempted and no-loss burglaries	51	6.1
Theft from cars	44	4.2
Common assault	31	5.0
Theft from the person	31	5.0
Poorly-reported crimes		
Theft in a dwelling	23	4.8
Vandalism over £20 to cars	30	4.5
Vandalism under £21 to the home	19	3.8
Sexual offences	10	9.1
Vandalism under £21 to cars	9	2.6

Notes

1. Question on seriousness: 'On this card is a scale to show the seriousness of different crimes with the scale going from 0 for a very minor crime like theft of milk bottles from a doorstep to 20 for the most serious crime, murder. How would you rate this crime on the scale from 0 to 20?
2. Weighted data: unweighted n = 3425.

Source

British Crime Survey 1984.

by victims' descriptions.

One likely reason for the shortfall is that the police do not accept victims' accounts of incidents: they may - quite rightly - think that a report of an incident is mistaken or disingenuous or may feel that there is simply insufficient evidence to say that a crime has been committed. Some incidents will have been recorded, of course, but in different crime categories - where, for example, it is indisputable that criminal damage has been committed, but less clear that a burglary has been attempted. Some incidents may have been regarded as too trivial to warrant formal police action - particularly if complainants indicated they wanted the matter dropped, or were unlikely to give evidence, or if the incident had already been satisfactorily resolved.

Three offence categories show no shortfall at all: sexual offences, thefts of motor vehicles and theft in a dwelling; in fact the BCS estimates of reported crime are lower than the number of recorded crimes. As the BCS estimate for sexual offences is unreliable little can be said about the discrepancy here. For thefts of motor vehicles the discrepancy was small and may have arisen simply through BCS measurement error. For thefts in a dwelling, the discrepancy was large and may be accounted for by differences in classification: some of the incidents which the BCS would classify as burglary may have been judged by the police to be the lesser offence of theft in a dwelling, perhaps because there was insufficient evidence of trespass.

Changes in crime over time

Now that two sweeps of the BCS have been carried out it is possible to say a little about movement in crime rates although what can be said on the basis of only two surveys is, of course, somewhat limited. Because of the sampling errors attached to victimization rates for the respective years only large changes in the rates will register as statistically significant. Even changes which are statistically significant do not necessarily reflect changes in underlying reality: they could arise, for example, from unintentional changes in survey procedure.

As shown in Table 2.2 between 1981 and 1983 there was a statistically significant rise of 10 per cent in the total number of household offences recorded in the BCS ($p<0.05$). However looking at household offences individually only the increase in burglary was statistically significant at the 5 per cent level, although the increase in bicycle thefts also approached this. The 21 per cent increase in burglary

Table 2.2 Offences in England and Wales, 1981 and 1983: British Crime Survey estimate

	1981	1983	% Change
1 Vandalism	2,714,000	2,953,000	+9
2 Theft from motor vehicle	1,272,000	1,364,000	+7
3 Burglary in a dwelling	745,000	904,000	+21**
4 Theft of motor vehicle	283,000	283,000	-
5 Bicycle theft	214,000	287,000	+34*
6 Theft in a dwelling	124,000	126,000	+2
7 Other household thefts	1,535,000	1,671,000	+9
8 Assaults	1,909,000	1,852,000	-3
9 Theft from person/robbery	596,000	650,000	+9
10 Sexual offences	33,000	71,000	(+115)
11 Other personal thefts	1,559,000	1,770,000	+14
All household offences (1-7)	6,887,000	7,588,000	+10**
All personal offences (8-11)	4,097,000	4,343,000	+6

Notes
1 Double-starred figures are statistically significant at the 5 per cent level (one-tailed test, taking complex standard error into account). This means that the chances are less than one in twenty that the increase has arisen simply through sampling error. Single-starred figures are statistically at the 10 per cent level. The increase in sexual offences will be due to questionnaire changes.
2 Categories 3, 7, 9, 10 and 11 include attempts.

reflects a sharp increase in the number of attempts: when these are excluded there is a much shallower increase of 11 per cent (ns: p>0.10). Part of the increase in bicycle thefts is probably attributable to increased ownership amongst adults and espcially amongst children. (The vogue for BMX bikes only emerged after 1981.) The 6 per cent increase in all personal offences was not statistically significant. (The increase in sexual offences is unreliable as explained because of deliberate questionnaire changes).

Do recorded crime figures show similar increases to those identified by the BCS? Limiting comparison to those crime types which can most reliably be compared the BCS shows a 10 per cent increase in crime - broadly in line with the 12 per cent increase in recorded crime figures. The divergence of 2 per cent is not statistically significant: in other words it could simply reflect sampling error. It might, however, also have arisen from increased reporting or recording. The former seems to be a likely explanation as the reporting levels identified in the two surveys increased between 1981 and 1983. Overall there was a (statistically significant $p<0.05$) rise in the proportion of offences reported to the police from 31 per cent to 34 per cent.

Comparison of survey estimates with police statistics is possible over a longer period for household burglary. Combining the GHS and the BCS, survey estimates of burglary rates can be calculated for the years 1972, 1973, 1979, 1980, 1981 and 1983. Figure 2.2 encapsulates the rationale for crime surveys very neatly. It compares survey estimates of burglaries involving loss to police statistics over the eleven-year period. The latter have doubled while the surveys indicate an increase of the order of 20 per cent. This divergence is sufficiently marked as to leave little doubt that police statistics exaggerate the increase in burglary rates since 1972 and that the differences between the two sets of figures can be explained by increased reporting to the police and increased recording by the police over this period.

Analysing crime risks: a conceptual model

Although crime surveys were originally conceived simply as social indicators it rapidly became clear that crime risks were not evenly distributed across the population but were heavily clustered amongst specific groups. The potential of crime surveys for describing who was at risk of crime soon became clear; and they have been the main data source used to support theories explaining the distribution of crime risks. Such theories have been variously labelled theories of lifestyle, opportunity, and routine activity.

Figure 2.2 Comparison of British Crime Survey estimates of burglary with police records over an eleven-year period

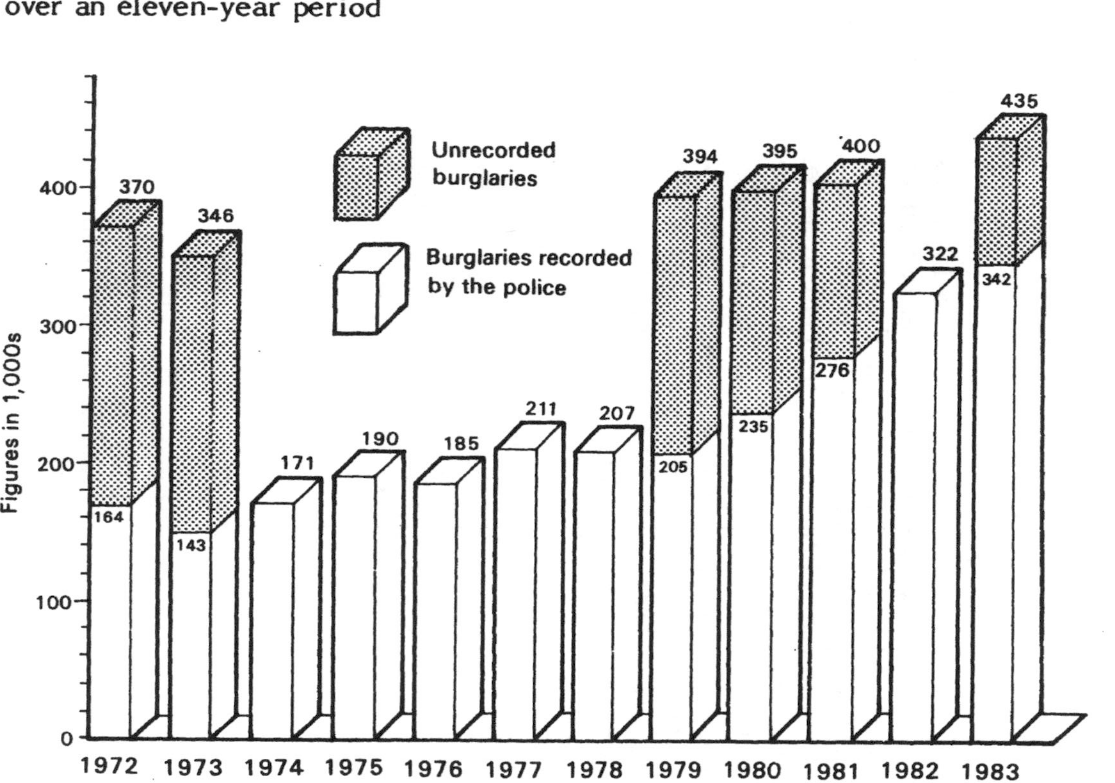

These kinds of theories are comparatively new to criminology. Traditionally the discipline has tended to focus on issues of motivation in explaining variations in crime, taking as constant (or irrelevant) changes in the structure of opportunities (see Clarke, 1984). Not surprisingly, therefore, the theories and concepts developed by criminology have been ill-suited to explaining variations between groups in victimization rates. However over the last decade researchers have begun to pay attention to opportunity - or from the victim's point of view exposure to risk - as a source of explanation. Notable examples include Mayhew et al. (1976, 1979), Hindelang et al. (1978), Steinmetz (1979), Cohen and Felson (1979), Cohen and Cantor (1981), van Dijk and Steinmetz (1983), Garofalo (1983), and Gottfredson (1984).

The basic axiom shared by these researchers is that for a crime to occur three minimal elements must converge in space and time:

- A motivated offender
- An attractive target
- An absence of protection or 'capable guardians'

The conceptual frameworks which they have developed are geared more to instrumental than expressive crime and have been concerned primarily with predatory crimes involving individuals and their personal property. For example, the 'routine activities' perspective developed by Felson, Cohen, and Cantor is particularly concerned to demonstrate how 'routine activities' alter the chances of victimization by altering the level of exposure to offenders. They defined routine activities as 'any recurrent and prevalent activities which provide for basic population and individual needs ... routine activities would include formalized work as well as the provision of standard food, shelter, sexual outlet, leisure, social interaction, learning and childrearing' (Cohen and Felson 1979). Similarly the 'lifestyle' model developed by Hindelang, Gottfredson, and Garofalo focused on people's day-to-day activities as a source of explanation of differing victimization risk - where 'lifestyle' is more or less synonymous with 'routine activities' (cf. Garofalo 1983). Both these approaches stress the importance of routine activities in determining crime risks but allow room for explanations of differential risk which turn on target attractiveness or accessibility.

Figure 2.3 A framework for analysing risk

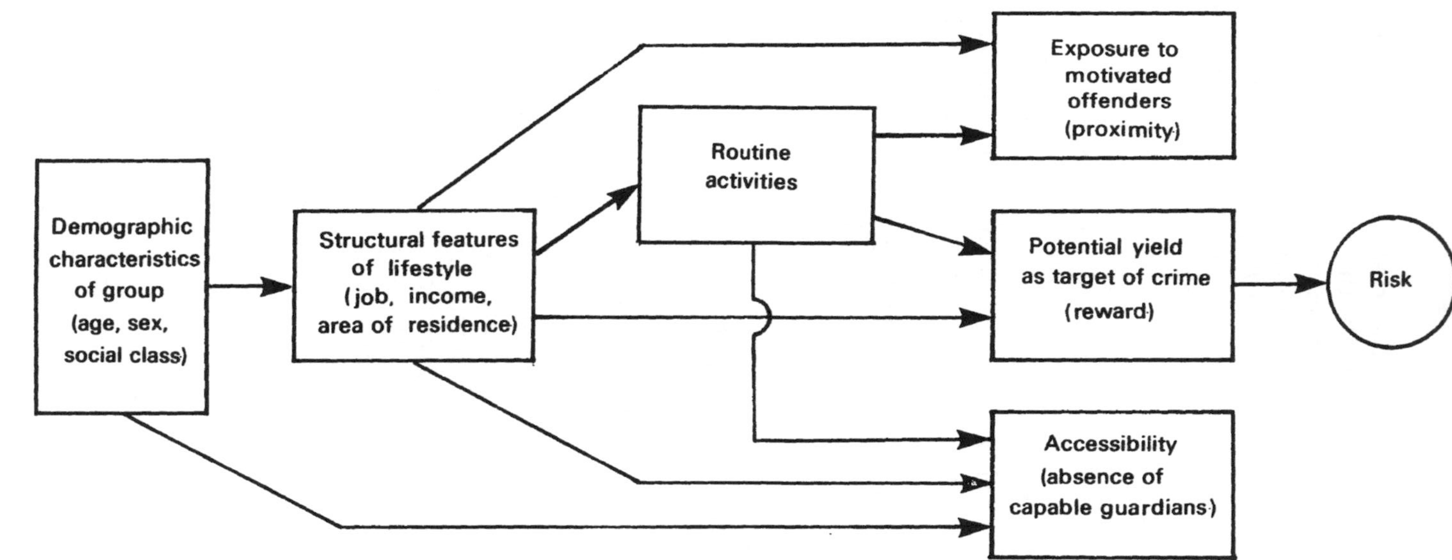

Van Dijk and Steinmetz (1983) have developed a very similar conceptual framework for analysing risk. They identify three risk factors: exposure or proximity to offenders, attractiveness, and vulnerability. They subdivide exposure into ecological proximity and exposure through lifestyle and subdivide vulnerability into 'technical' and 'social control' vulnerability. Figure 2.3 offers (yet another) variant of this conceptual framework.

This model takes it as axiomatic that if one group is more frequently victimized than another at least one of three conditions (or a mixture of them) must be being met: (i) its members must be exposed more frequently to motivated offenders; or (ii) be more attractive as targets in affording a better 'yield' to the offender; or (iii) be more attractive because they are more accessible or less defended against victimization. The framework bears a very close relationship to Garofalo's (1983) modified lifestyle model and to the framework offered by the Dutch researchers. In fact the differences are more of nomenclature than of substance and are largely a matter of taste - but four points are worth making here.

First, both Garofalo and Van Dijk and Steinmetz use the term 'attractiveness' to refer solely and specifically to the 'instrumental or symbolic worth' of the target to the offender: but targets can also be attractive not just for the rewards they offer but because they are easy targets. To maintain this distinction it is probably better to relabel attractiveness 'potential yield'.

Second, the term 'vulnerability' - often used in the past - is on the one hand tautologous (one could argue that vulnerable targets are those which get hit for whatever reason). On the other hand it can be misleading since victimologists are increasingly using this term to refer to the differential impact that an objective level of harm causes to one victim (or group of victims) over another. The model drawn here, therefore, uses 'accessibility' to refer to the ease of committing an offence that a target offers a potential offender. (Garofalo in his revised lifestyle model embraces the concept of accessibility in the broader concept of exposure to motivated offenders.)

Third, the terms 'routine activity' or 'lifestyle' - which can determine proximity to offenders, yield, and accessibility - are rather elastic. People can be exposed to motivated offenders by virtue of their routine activities: how they spend their free time, how they travel to and from work - those aspects of their lives which we generally regard as controllable. Equally exposure can be explained by more structural features of people's lives: the nature of their job, where they live, how much they earn, for example - which are less mutable and less under people's control. Are area of residence or patterns of home occupancy features of lifestyle? The term 'lifestyle' has voluntaristic connotations in everyday usage: we would probably say that the homes of the affluent professional classes are a reflection of their lifestyle but would baulk at the idea that ghettoes and occupations of the urban poor are characteristics of <u>their</u> lifestyle. In short, the distinction between lifestyle and socio-demographic characteristics is inevitably arbitrary.

Finally, operationalizing concepts of exposure, yield and accessibility is a real problem because of the difficulty of finding quantitative measures which adequately reflect the concepts involved. The concepts of exposure, yield and accessibility may be conceptually distinct but in the real world their relationships are often hopelessly intertwined. Take a man walking home late at night in the inner city - an archetypal prospect for robbery. Make him drunk: this will simultaneously affect his exposure to offenders (they were also drinking in the pub), his accessibility (a push-over), and possibly the reward he offers (the race-course winnings which set him on his spree to start with). And besides this, of course, very few data sources combine information on victimization with detailed information on routine activities and lifestyle. Too often, researchers are reduced to using demographic variables as proxies for lifestyle.

An example: the risks of burglary

The two sweeps of the BCS carried out in 1982 and 1984 amount to a fairly extensive data-set for examining factors shaping burglars' choice of target. The rest of this paper presents some of the findings on burglary risks from the two surveys.

Proximity

Both sweeps of the BCS have shown that for most offence categories crime is an urban phenomenon and that rates are highest in inner-city areas. The explanatory power of area

Table 2.3 Burglary risks and fears by Acorn neighbourhood group, 1983

	HHs burgled (including attempts) in 1983 %	HHs burgled (excluding attempts) in 1983 %
Low-risk areas		
A Agricultural areas (n = 476)	1	1
C Older housing of intermediate status (n=2001)	2	1
K Better-off retirement areas (n=463)	3	1
J Affluent suburban housing (n=1659)	3	2
B Modern family housing higher incomes (n=1537)	3	2
Medium-risk areas		
E Better-off council estates (n=1018)	4	2
D Poor-quality older terraced housing (n=759)	4	3
F Less well-off council estates (n=1175)	4	2
High-risk areas		
I High status non-family areas (n=609)	10	6
H Multi-racial areas (n=400)	10	6
G Poorest council estates (n=543)	12	7
National average	4	2

Notes
1 Weighted data: the numbers of each Acorn group are unweighted.

Source
British Crime Survey 1984.

Table 2.4 Percentages of households burgled by household income and area

	Below average income % burgled	Average income % burgled	Above average income % burgled
Rural areas	2	2	3
Towns	3	4	6
Cities	4	4	8
Inner cities	7	8	16
Total	4	4	7

Notes
Weighted data
Unweighted n = 7986

Source
British Crime Survey 1984.

variables outstrips that of most others for most crimes. Table 2.3 shows burglary rates in England and Wales by Acorn classification: variations are very marked with especially high rates in three types of area, all associated with inner cities (or the 'overspill' estates on the perimeter of conurbations).

Though there is no evidence internal to the BCS to prove it, (1) it is reasonable to assume that risks are high in these areas because they contain the greatest density of offenders. What evidence there is about the mobility of burglars suggests that most - especially juveniles - operate within a short distance of their homes (e.g. Baldwin and Bottoms 1976). Presumably this is because they are too young to drive or cannot afford cars and are unprepared to travel long distances on public transport with stolen goods.

Potential yield

How important is the prospective yield of targets to burglars in choosing targets? Table 2.4 shows that poverty provides virtually no protection: poor homes are burgled only marginally less than those with average incomes. The better-off are more vulnerable, however. These results

Table 2.5 Day-time and night-time burglary rates by occupancy

Day-time burglary rates by occupancy

Occupancy	Day-time rate
Home empty day	%
Not at all	0.6
Not often	1.1
Very often	1.9

Night-time burglary rates by occupancy

Occupancy	Night-time rate
Home empty evening	%
Not at all	1.8
Not often	1.5
Very often	2.8

Source
British Crime Survey 1982.

could be explained in terms of the different styles of operation of skilled and unskilled burglars. The former presumably invest considerable effort in locating profitable targets - sometimes travelling long distances to reach them. The less skilled, by contrast, are more likely to tackle available targets indiscriminately - explaining, perhaps, why the Acorn area-type with highest burglary rates are poorest council estates and why within these estates risk seems to be unrelated to income.

Accessibility

In the case of burglary, accessibility or the absence of 'capable guardians' can be broken down into several constituents. Only a small proportion of the burglaries reported in the BCS had been successfully completed when householders were at home. The presence of residents themselves thus provides considerable protection. Table 2.5 shows how patterns of occupancy affect burglary rates at different times of day.

Burglars obviously take into account factors besides occupancy. The first BCS showed that physical accessibility

Table 2.6 Percentage of households burgled by accessibility and area

	Easy access to back of the home % burgled	Less easy access to back % burgled
Rural areas	2	2
Towns	4	2
Cities	6	5
Inner cities	16	11
Total	4	3

Notes
Weighted data.
Unweighted n = 5481.

Source
British Crime Survey 1982.

to the rear of properties is important: houses at the end of terraces, for example, were more frequently chosen than those in mid-terrace and as Table 2.6 shows homes with easy rear access were more vulnerable.

As for other aspects of accessibility the main lesson from the BCS is about measurement problems. The first BCS tried to measure the extent of 'natural surveillance' protection afforded by neighbours and passers-by, by asking how overlooked respondents' homes were. But the responses were heavily skewed with three-quarters of all homes scoring high on surveillance and the results are difficult to interpret (see Hope 1984). Surprisingly the survey failed to show that dog-ownership conferred any protection on homes despite burglars' claims that they avoid dogs. Again this could reflect measurement difficulties: we asked about dog ownership at the time of interview and a proportion of victims may have bought dogs for protection after their burglary. Similarly the BCS asked about use of security hardware at the time of interview and failed to find any relationship between security and vulnerability - presumably because no one is more security-conscious than a recent victim of burglary.

Summary

This chapter has discussed the rationale of surveys of victimization or crime surveys and has described the design and main findings of the British Crime Survey: it went on to examine the analysis of crime risks using burglary as a 'worked example'.

The main aim of the BCS was to provide an estimate of rates of victimization to supplement police statistics. By asking victims directly about the crimes they have suffered it is possible to gain information both about offences which have been reported to the police and those which have not and to look in more detail at the kinds of factors which prompt victims to report incidents. Analysis of the BCS shows, as might be expected, that in most cases the severity of the offence is the crucial factor in decisions to report.

On crime trends, police statistics and the BCS showed similar rates of increase between 1981 and 1983 though there was some evidence of increased reporting to the police. Combining BCS data with findings from the General Household Survey there is clear evidence that the increase in burglary between 1972 and 1983 reflects in part increased reporting to the police and increased recording by the police. The third sweep of the BCS in 1988 should throw more light on trends in the reporting and recording of offences.

The chapter outlined a conceptual framework for examining risks of crime using data from the BCS on household burglary. It illustrated some of the ways in which proximity, accessibility, and potential yield seem to affect offenders' choices of targets and hence probabilities of victimization. Future sweeps of the BCS will be able to provide more of this kind of data to help identify factors relating to risks and possible ways in which these can be tackled.

Notes

1. Several self-report offending questions were asked in the 1982 survey but these proved of doubtful reliability: the item on burglary showed no variation across area.

References

Baldwin, J. and Bottoms, A.E. (1976) The Urban Criminal, London: Tavistock

Chambers, G. and Tombs, J. (eds) (1984) The British Crime Survey: Scotland, Scottish Office Social Research Study, Edinburgh: HMSO

Clarke, R.V.G. (1984) 'Opportunity-based crime rates', British Journal of Criminology 24, 1, 74-83

Cohen, L.E. and Cantor, D. (1981) 'Residential burglary in the United States: life-style and demographic factor associated with the probability of victimization', Journal of Research in Crime and Delinquency, January, 113-27

Cohen, L.E. and Felson, M. (1979) 'Social change and crime rate trends: a routine activity approach', American Sociological Review, 44, 580-608

Farrington, D.P. and Dowds, E.A. (1985) 'Disentangling criminal behaviour and police reaction', in Farrington, D.P. and Gunn, J. (eds), Reaction to Crime: the public, the police, courts, and prisons, Chichester: John Wiley

Garofalo, J. (1983) Lifestyles and victimization: an update, Paper presented at the 33rd International Course in Criminology, Vancouver, BC, Canada, March 1983

Gottfredson, M.R. (1984) Victims of Crime: the dimensions of risk. Home Office Research Study No. 81, London: HMSO

Hindelang, M.J., Gottfredson, M.R., and Garofalo, J. (1978) Victims of Personal Crime: an Empirical Foundation for a Theory of Personal Victimization, Cambridge, Mass.: Ballinger

Hope, T. (1984) 'Building design and burglary', in Clarke, R.V.G., and Hope, T. (eds), Coping with Burglary, Boston, Mass.: Kluwer-Nijhoff

Hough, M. and Mayhew, P. (1983) The British Crime Survey: First Report, Home Office Research Study No. 76, London: HMSO

Hough, M. and Mayhew, P. (1985) Taking Account of Crime: Key Findings from the 1984 British Crime Survey, Home Office Research Study No. 85, London: HMSO

Jones, T., Maclean, B., and Young, J. (1986) The Islington Crime Survey, Aldershot: Gower

Kinsey, R. Merseyside Crime Survey: First Report (1984), unpublished report to Merseyside County Council, Centre for Criminology, University of Edinburgh

Mayhew, P., Clarke, R.V.G., Sturman, A., and Hough, J.M. (1976) Crime As Opportunity, Home Office Research Study No. 34, London: HMSO

Mayhew, P., Clarke, R.V.G., Burrows, J.N., Hough, J.M. and Winchester, S.W.C. (1979) Crime in Public View, Home Office Research Study No. 49, London: HMSO

NOP Market Research Limited. (1985) 1984 British Crime Survey Technical Report, NOP/9888, London: NOP Market Research Limited

Skogan, W.G. (1981) Issues in the Measurement of Victimization, Bureau of Justice Statistics, NCJ-74682, Washington DC: US Department of Justice

Sparks, R.F. (1982) Research on Victims of Crime: Accomplishments, Issues, and New Directions, National Institute of Mental Health, Rockville MD: Centre for Studies of Crime and Delinquency

Steinmetz, C. (1979). An Empirically-Tested Analysis of Victimization Risks, Report No. XXIV, The Hague: Ministry of Justice, Research and Documentation Centre

van Dijk, J.J.M. and Steinmetz, C. (1983) 'Victimization surveys: beyond measuring the volume of crime', Victimology 8, 291-301

Wood, D.S. (1984) British Crime Survey: Technical Report, London: Social and Community Planning Research

Chapter Three

ASSAULT AND HEAT STRESS: DALLAS AS A CASE STUDY

Keith D. Harries and Stephen J. Stadler

Introduction

Assault was selected for study for several reasons. First, it is endemic in the US and it accounts for a great amount of human suffering. Understandably, people generally perceive crimes that threaten or terminate life to be among the most heinous. The US National Survey of Crime Severity found that people assigned the highest seriousness scores to events involving violence or drugs (Bureau of Justice Statistics 1983). Some 723,000 Americans were reportedly victims of aggravated assault in 1985, signifying vast aggregate medical costs, as well as untold psychological damage, and staggering losses of earnings (Federal Bureau of Investigation 1986).

Levels of violence found in American cities might well engender a state of emergency in Europe or other developed areas. If homicide can be used as a representative measure of violence, we can compare Northern Ireland with the US. Northern Ireland has about 1.5 million people, roughly comparable to the City of Philadelphia. Yet the average annual loss of life in Northern Ireland since the late 1960s (about 120) gives it a homicide rate of some 8 per 100,000, less than half the current rate in Philadelphia. The City of Brotherly Love has a police force of about 8,000 compared to the more than 27,000 police and British Army troops in Northern Ireland. This provides an interesting commentary on norms; over 300 homicides a year in Philadelphia are to be expected. That many in Northern Ireland would presumably lead to an even more intense level of the deployment of military personnel and police to maintain law and order.

Second, aggravated assault, defined as an 'unlawful attack by one person upon another for the purpose of inflicting severe or aggravated bodily injury' (Federal Bureau of Investigation 1986) is relatively well reported. National Crime Survey data found that aggravated assault was reported above the 60 per cent level, compared to 51 per cent for household burglary and 26 per cent for household larceny (Bureau of Justice Statistics 1983). Thus it is likely that the incomplete report data represent a particular class of criminal behaviour reasonably well, even taking into account spatial biases in reporting, which probably under-represent high rate areas.

Third, assault was preferable to homicide for the original purpose of our study, in that the other possible crimes of violence: rape, homicide, and robbery, were contaminated by various inadequacies. Even in large cities, homicide exhibits too few events to permit fine-grained temporal analysis. Rape is notoriously under-reported. In Dallas it had only about one sixth the frequency of aggravated assault, again rendering it of dubious value for the purpose of temporal analysis. Robbery, although resulting in many homicides and assaults, is primarily an instrumental offence and would not have captured the expressive behaviour component that we wished to analyse.

Conceptualization

Most research relating to violence has been oriented towards the individual level of analysis (Sampson 1986), and the focus has tended to be on the offender rather than the victim (Braucht et al. 1980). Although the role of environmental factors has attracted more attention in recent years, a balanced view of all psychological, social, economic, and physical environmental variables has yet to be achieved, and may not be possible. This is due in part to the difficulty of reaching consensus on the nature of interactions in and between each subset. Arguments about the role of economic inequality in homicide exemplify this debate (see Loftin and Parker 1985). While consensus and balance are elusive, social ecology has tended to stress the importance of several clusters of variables in the search for explanation: family environment, peer environment, and situational factors (Gabor 1986). More recent developments in environmental criminology have a strong focus on the last of these and the places at which criminal events occur.

Family environment stresses the role of the family in the transmission of values and as the fundamental unit of social interaction. The peer environment is the context in which youths, in particular, learn and derive their inspiration for specific behaviours. Situational factors are dealt with in more detail by Davidson in Chapter 4. They include the social, demographic, and economic characteristics of the area under analysis, the distribution of high-risk sites, the level of interaction with those sites, the relationship between criminal opportunities and probabilities of detection, considerations of territoriality, and the substance abuse environment. These factors are set in a context of place, space, and location. Individuals are nominally able to control some aspects of their environment, such as who they choose to associate with and where they choose to be (at least for some activities). Other aspects of environment are not amenable to control (family, race, age).

Weather is a situational factor that has received very little attention for a variety of reasons. First, as indicated more fully below, theoretical links between weather and criminal behaviour have been speculative and have been associated with the excesses of what might be termed 'Huntingtonian' environmentalism (see p. 43-4). Second, weather conditions vary on a regional basis and are insufficiently extreme in many areas to warrant the suggestion that deviant behaviour could be explained to any useful degree by atmosphere anomalies. Even in areas experiencing weather extremes, some years are much less extreme than others and not all years are equally suitable for the study of weather as a situational factor. Third, the concept of weather as 'stress' has been neglected in the field of stress research, which has dwelt on illness and disease as products of stress. As Linsky and Strauss (1986) have noted, stress research is rooted in psychosomatic medicine and mental health, rather than criminology. However, the stress perspective provides a useful bridge between 'conventional' situational factors used in criminological studies, quality of life (QOL) analyses, and medical perspectives now coming into vogue in the US. An example can be seen in Persinger (1980) where weather and human behaviour are treated as parts of the same matrix.

Linsky and Strauss (1986) suggest several channels

through which stress may operate to engender criminal activity. First, is the 'fight or flight' reaction to threatening situations, a physiological response caused by changes in adrenal secretions. Second, interaction between social control and stress may lead to disease (when social control is strong and physiological arousal is constant), or to behavioural reaction in the form of suicide or depression. When social control is weak, aggression may be the direct product of arousal. Third, stresses may be legitimized over time by the culture of a group, as in the case of the Southern violence phenomenon in the US (Harries 1985). Fourth, the process of change and destabilization in the course of such events as divorce and migration has the effect of stressing individuals and removing informal social controls. Fifth, law enforcement may be affected by the cumulation of stressful events, in that it may react selectively to groups or situations that have become the focus of public concern, such as the 'mod and rocker' phenomenon in the UK in the 1960s (Cohen 1980).

Linsky and Strauss (1986, 77) felt able to conclude that: 'stressful life events are positively correlated with all seven of the so-called 'index crimes' ... cumulative stressful events are causally related to each of these crimes'. Their analysis of aggravated assault produced a regression coefficient of 7.39, indicating an increase of >7 assaults per 100,000 population for each one point increase in the stress index.

Whether stress theory and the more familiar QOL concept are significantly different as they relate to crime is moot (cf. Smith 1973; Harries 1976). It would appear that high stress and low QOL are strongly correlated. Indeed, some of the same variables are typically used in the construction of both stress and QOL indexes. The proponents of stress theory emphasize the importance of cumulative effects. Again, it would seem that the concept of cumulation or additive effects is implicit in QOL. However, stress theory may have a more coherent theoretical base than QOL analyses in that stress may be more amenable to measurement and conceptualization than 'quality'.

In this vein, our focus on heat stress is congruent with other aspects of stress theory. There is no question that the average person is physiologically stressed under certain combinations of heat, humidity, and air movement. Although our work concerns heat stress, it has been suggested that comparable investigation should also be directed toward 'cold stress', which involves not only the physiological effects of cold, but also the proverbial 'cabin fever' suffered by those who are confined to dwellings or workplaces by prolonged periods of extreme cold. In the US, this phenomenon is most pervasive in the Upper Midwest - Minnesota, Wisconsin, Michigan, for example - where winters are long and severe, with daily maxima at or below -18°C (0°F) commonplace. From a criminological perspective, a derivative of cabin fever is increased actual or potential interpersonal interaction, and hence potential violence. Although weather-induced stresses with the potential to contribute to deviant behaviour have been recognized in folklore for millennia, they have been almost totally ignored as components in the matrix of social science research.

Background

Two thousand years ago the Greeks produced a substantial body of writings stemming from the idea that weather and climate affect human behaviour. Glacken (1967) traced parallel environmental theories: physiologic and geographic. The physiologic approach focused on relationships between the human organism and environmental conditions; important contemporary works such as Landsberg (1969) and Tromp (1986) draw on the intellectual heritage of the Greeks. The other part of the Greek legacy, the geographic approach, deals with variations in the environment and their consequences and has periodically undergone severe criticism.

If the root of physiologic environmental theory is to be found in the works of Hippocrates, so too is the basis of geographic environmentalism. Hippocrates admonished physicians to first study the atmosphere when taking up residence in a new city; he believed that diseases varied with climatic differences. At the time of the development and application of scientific instruments, both the physiologic and geographic species of environmental theory were alive. However, instrumentation tended to favour physiologic theory in that whereas physiologic relationships could be readily demonstrated through direct observations of individuals, geographic relationships could not.

Despite difficulties in the production of plausible geographic environmental theory, some writers did foster the notion of a link between temperatures and behaviour. In the 1700s, the Abbé du Bos attributed criminal behaviour to abnormally high Roman summertime temperatures (Koller 1937). Kant believed that too much heat dried out nerves and veins (Oliver and Siddiqi 1987). Geographic environmental theory though, also had powerful detractors such as Voltaire who discounted any deterministic effects of the atmosphere (Glacken 1967).

With the publication of On The Origin of Species in 1859, geographic environmental theory was stimulated, particularly after the transposition to Social Darwinism. Stoddart has argued that it reached all corners of geography and that geographers received the ideas with some fervour. He noted that study of the physiological effects of the environment increasingly grew 'outside geographical competence' and that 'geographers have generally restricted themselves to the inference of causation on a coarser scale' (Stoddart 1972, 62). By its very nature in attempting to straddle the social and natural sciences, geography was relegated to explanation of environmental effects upon society with broad, sometimes unsupported, brush strokes.

In the language of Social Darwinism, Ratzel's Anthropogeographie (1882) theorized that the direct and indirect effects of climate were powerful influences upon human behaviour. His student, Semple, popularized his ideas in the fledgling graduate geography departments of the US. Semple's literary style was persuasive, though James and Martin (1985, 306) have argued that she had a propensity to 'carry the theme beyond what sober judgement would permit'.

Early academic geography in the United States was steeped in this climatic determinism, with Ellsworth Huntington as the most influential representative. Although his work is remembered with general disdain, he had a profound influence in his day. Books such as Civilization and Climate (1915), Mainsprings of Civilization (1945), and The Pulse of Asia (1907) were broad treatises attempting to link geographical differences in climate with overviews of differences in civilizations. His writings enjoyed a reputation far outside academia and some of his contemporaries were persuaded by his success. However, geographers increasingly opposed determinism in reaction to publications such as Climate Makes the Man (Mills 1942) and the works of Griffith Taylor (1955), who viewed human geographies as being controlled by climate. Taylor argued that the climatic environment provided bounds which should be observed by humans; he was in part vindicated by his celebrated prediction that the Australian outback would not become densely settled and Australia even issued a stamp in his honour (Holt-Jensen 1980). The historian Toynbee, too, carried Huntington's banner with the sweeping and influential series A Study of History (1955). Ironically, the downfall of the determinists lay in their attempts to integrate disparate threads of geography and history. The most common critique has been that they employed only the facts and impressions which supported their contentions and excluded equally viable, but contrary, evidence (see Oliver 1973; Lee 1954; Spate 1952). The analytical inadequacies of the climatic determinists were due as much to the lack of reliable sources and computational capabilities as to Zeitgeist. Moreover, when the intellectual ties between geographic environmental theory and climatic determinism are examined, it is clear that they became one in the early twentieth century. Discrediting the climatic determinists served to halt geographic environmental theory in its tracks. In essence, threads of two millennia of environmental thought had been obliterated by the time climatic determinism had run its course.

Geographers have been understandably reluctant to pick up the debris of this environmental tradition. Rejection of determinism was reinforced by the emergence of socially-rooted topics important in their own right (Martin 1973). The reaction to determinism was so strong that by 1954 Lee could write with assurance that physiological climatology was progressing and that Huntingtonian determinism was all but dead (Lee 1954).

Although folklore, experience, and common observation demonstrate links between climate and behaviour, such research in geography became virtually moribund. In the 1960s and early 1970s, a little research of a nominally environmental turn was done by geographers, but it was a cadre of pyschologists who initiated studies at both the macro and clinical levels, looking specifically at crime-weather relationships. These psychological studies have provided a reasonable link between abnormally high temperatures, irritability, and violent tendencies in laboratory subjects (Boyanowski et al 1981; Baron and Bell 1975; Sutherland and Cressey 1978; Baron 1972). Although psychology has harboured much controversy (Rotton 1986), clinical studies support the notion that violence might be partially explained through high temperatures.

Another theme identifiable in environmental research is the relationship between seasonality and crime. Every study of temperature and violence has shown a warm-season peak. Guerry recorded a summertime zenith in France in the 1830s (Anonymous 1833). Dexter used an eight-year time series to connect temperatures and criminal assault in turn-of-the-century New York City (Dexter 1904), and Cohen's examination of the Uniform Crime Reports of the 1930s showed the same phenomenon on a national scale (Cohen 1941). Nineteen-eighty saw the publication of Crime and Seasonality which amply demonstrated warm-season peaking in crime (Bureau of Justice Statistics 1980). Yet the interpretation attributed the peaking entirely to non-thermal factors such as summer school vacations, holidays, and 31-day summer months. Geographers Lewis and Alford also took a seasonal perspective when they attempted to find climatic differences in the timing of the onset of higher warm season assault rates in the United States (Lewis and Alford 1975). They theorized that the South should be the first region in the calendar year to experience higher assault rates and suggested that this might be associated with critical temperature thresholds. However, using national data, they found no tenable relationship.

The contemporary approach to geographic environmentalism, illustrated here, recognizes that environmental influences are complex and often marginal, but may be strong enough to warrant recognition and further exploration.

Summary of recent findings

We have undertaken three studies that have viewed the issue of thermal stress and violence in Dallas from somewhat different perspectives. The theoretical approach in this research rests on findings from a combination of folklore, clinical and social psychology, physiological climatology, and the sciences of indoor climate control and clothing. Expressed in the simplest terms, we suggest that violence is a product of various complex environmental interactions, including the innate characteristics of the actors, their social context, and the physical environment, of which the atmosphere is a part. We recognize that atmospheric influences are typically small, but their existence should be recognized and the effects estimated as far as possible.

The immediate theoretical precedent for the research came from clinical psychology, where evidence had accumulated to suggest that subjects stressed thermally became irritable and more prone to violence compared to a control group (Boyanowski et al 1981; Calvert-Boyanowski et al 1976). If thermal stress is indeed a universal phenomenon, then a relationship with officially measured violent behaviour should be detectable in thermally stressful environments. For the purpose of investigating the hypothesized effect, the ideal environment would have to exhibit extreme conditions over a protracted time period. Furthermore, local levels of 'normal' violence would have to be rather high in order to provide a sufficient number of incidents to allow some statistical insights and permit some degree of disaggregation of the data. Dallas, Texas, qualified according to these criteria.

The data base for all three studies consisted of assault data for the period March through October, 1980, and meteorological conditions for the same period. The study period was selected in order to permit the 'bracketing' of the warmest months in order to ensure the inclusion of the most severe summer weather. Nineteen-eighty was selected for the extraordinary severity of its summer heat. The discomfort was particularly striking since it was conspicuous in an environment of climatic extremes. People in the southern Great Plains are accustomed to an annual temperature range of the order of at least 38°C (100°F), thunderstorms, drought, and tornadic winds. At the Dallas-Fort Worth airport, located at the fringe of the urban heat island, a temperature of 46°C (115°F) was recorded in 1980. It is likely that ambient shade temperatures in some parts of the inner city exceeded 49°C (120°F). Media attention focused on Dallas and other cities similarly affected in the South West, drawing attention to high rates of heat stroke and deaths attributable to heat stroke, particularly among the elderly. Media accounts tended to dwell on the various stresses caused by the heat, and their consequences. Emergency efforts attempted to distribute electric fans to those without air conditioning.

During the eight-month study period, some 4,309 aggravated assaults were reported to the police in the city of Dallas. Data on these incidents were acquired from the Dallas Police Department. Undoubtedly, the count of assaults was an under-enumeration of the actual incidence, as noted earlier. However, the nature of under-enumeration is reasonably well understood, and it is generally accepted that communities or neighbourhoods of higher socio-economic status are more likely to report compared to lower status areas, where crimes of all sorts, including violence, are commonplace. The three studies dealt with:

(i) The relationship between assaultive violence and a widely used discomfort index (DI) (Harries and Stadler 1983)
(ii) Differences in neighbourhood responses to thermal stress (Harries, Stadler, and Zdorkowski 1984)
(iii) A hypothesized threshold effect, suggesting that violence will actually diminish above some temperature or comfort level, beyond which people become so uncomfortable that virtually all activities, including violence, are inhibited (Harries and Stadler in press)

Assault and discomfort

Only a few studies in geography (Lewis and Alford 1975) and social psychology (e.g. Baron and Ransberger 1978; Carlsmith and Anderson 1979) had considered the question of temperature effects as they related to violence. This research had either (i) not considered temperature directly, but had relied, rather, on a seasonal surrogate (Lewis and Alford), or (ii) had used ambient temperature as the measure of thermal stress. Such approaches are either too coarse, in that they cannot hope to capture the possibly subtle relationships between thermal stress and the incidence of violence, or may be weak measures of discomfort, as in the use of ambient temperature.

It seemed appropriate, then, to test a DI taking into account not only temperature but also humidity, given the importance of the interaction between these conditions in the calculus of comfort. Numerous DIs have been proposed, varying substantially in practicability. The 'best' measures in terms of precise description of human discomfort are impractical in epidemiological studies. The three-dimensional patterns of urban thermal discomfort are quite complex and a complete description would have to account for the myriad combinations of temperature, radiant load, air movement, and humidity. A city-wide study must rely on some surrogate measures of thermal discomfort. In the case of Dallas, we used data from the Dallas-Fort Worth regional airport located at the urban fringe. As suggested above, the level of discomfort was probably underestimated by both the ambient temperature and DI values. The 4,309 incidents were aggregated into 245 daily assault counts, and the DI was computed from nearly 6,000 hourly weather observations obtained from the National Climate Center. Daily maximum ambient temperature was also abstracted for comparative purposes.

Assault is by definition a human interaction. As such, it is most likely to occur in informal social settings in homes or places of recreation. Such violence is apparently less likely in the workplace owing to the inhibiting effect of the public or quasi-public nature of most work settings, combined with the immediate threat of job loss. Assault, therefore, has a distinct weekly rhythm, with weekend peaks and weekday lows. Given that opportunities for some types of recreation are increased in the summer owing to extended daylight hours and warm weather, a seasonal peak might be expected in the summer. However, a counter-argument can be developed to the effect that social interactions may be most intense in winter when people are confined in indoor settings for prolonged periods.

We developed an analysis of variance model using day of the week, month (to capture possible seasonal effects), and DI as explanatory variables. The raw DI values were converted to a five-level classification reflecting generalized human reactions to the conditions represented by the index. The thresholds for these levels were DIs of 65, 75, 80, and 85. The literature has established that virtually all people are comfortable at DIs below 65; above 85, most people experience severe distress. The three variables were quite successful in accounting for variations in assault frequencies ($F=3.74$, $p<0.0001$), with $R^2 = 0.71$. At the lowest discomfort level, some 15.7 assaults per day occurred; at the level of most discomfort the mean assault frequency was 20.3. The DI was still significant when it was forced last in the model ($p = 0.05$). The overwhelming effect of day of the week was illustrated by mean assault frequencies for Wednesdays (13.5) and Saturdays (27.0).

This analysis suggested that weather effects, almost entirely overlooked by criminologists and others in causal studies of violence, should be considered, particularly in environments prone to severe conditions.

Neighbourhood response

Having demonstrated a general thermal stress effect, our attention moved to intra-urban geography in order to examine the idea that neighbourhoods classified on the basis of socio-economic status (SES) could be expected to respond differentially to thermal stress. The rationale was that the affluent, through employment in indoor occupations, and air conditioning in cars and homes, would be able to ameliorate the effects of thermal stress. The poor, on the other hand, would be more likely to be employed in outdoor settings, and less likely to have access to functioning air conditioning.

Census data showed that about 90 per cent of the dwellings in Dallas were air-conditioned in 1980. However, over 95 per cent of the homes occupied by whites were cooled, compared to 83 per cent in black-occupied dwellings and 79 per cent in homes occupied by Hispanics. The proportion of air conditioning systems that was central (one relatively efficient unit as opposed to smaller, less efficient units in one or more rooms) showed similar variation. Some 37,000 dwelling units in Dallas had no air conditioning in 1980. Based on the assumption that each dwelling was occupied by the average number of persons per household in Dallas in 1980 (2.51), it is estimated that some 93,000 persons were living without air conditioning. Based on the geography of low SES neighbourhoods in the metropolitan area, it seemed that most of the people without air conditioning were likely to be subjected to the most uncomfortable temperatures of the urban heat island. According to the stress model, the lower SES people, poorly equipped to cope, could be expected to respond either behaviourally or in the form of mental or physical illness, or both.

Neighbourhood status was measured by an Urban Pathology Index constructed for neighbourhood clusters and based on rankings of sub-standard housing, black population, and median household income. The index discriminated clearly between three groups of neighbourhoods, designated as high, medium, and low status. Other factors, including the urban heat island, population density, substance abuse, and 'calendar effects' should also be considered as potential contributors to violence, directly or indirectly.

Our data did not permit intra-urban analysis of thermal effects, and the heat island hypothesis was not tested. Population density could contribute to violence in that it is a surrogate for potential social interaction. Data relating to

Figure 3.1 Relationship between assault frequencies, months, and neighbourhood status, Dallas, 1980

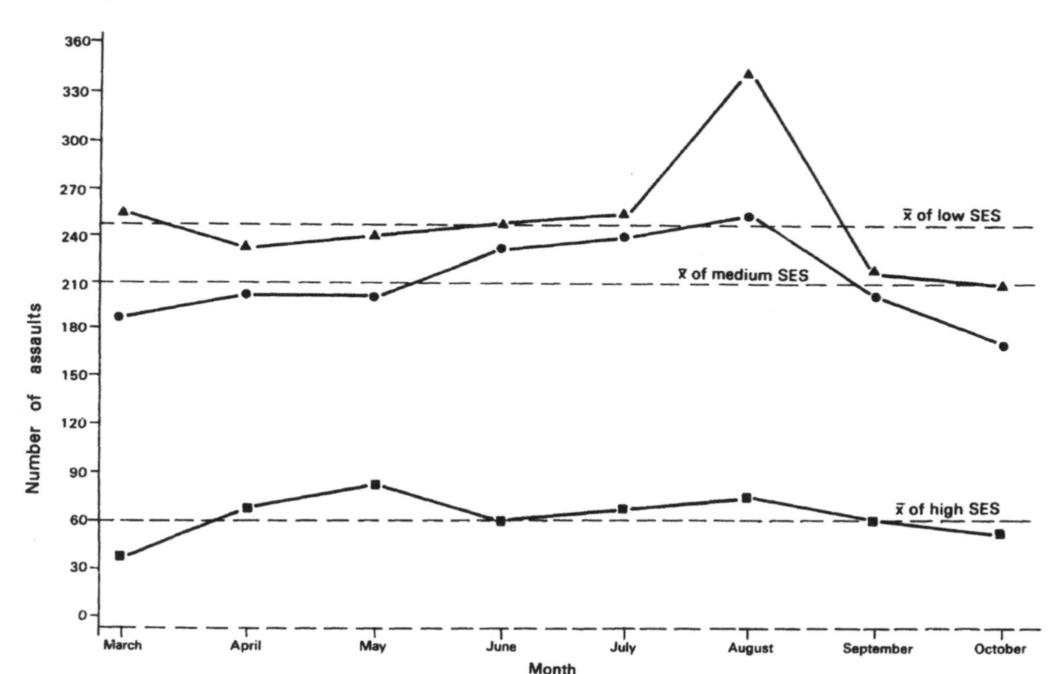

Source: authors. (Copyright Association of American Geographers, 1984. Reproduced by permission.)

persons in multi-family dwellings provided a crude indication of density. Substance abuse could not be measured satisfactorily, but some indication of the role of alcohol was provided by counts of the number of incidents occurring in bars or in bar parking lots. 'Calendar effects' refers to variations in activities caused by events with somewhat predictable schedules. These included weekends, school holidays and graduations, various public holidays, paydays, sporting events, and (some would argue) lunar cycles. The latter issue, apparently, can be dismissed (see Rotton and Kelly 1985; Harries 1986). The calendar effects hypothesis is akin to what has come to be known in criminology as the 'routine activities' approach (Cohen and Felson 1979; Messner and Tardiff 1985). However, a detailed investigation of the calendar effects hypothesis, like the urban heat island, remains on our research agenda.

Neighbourhood SES level was strongly and inversely related to assault incidence. The expected exaggerated response of low SES neighbourhoods was confirmed; August exhibited a sharp peak, but for the low SES areas only (Figure 3.1). The density hypothesis was evaluated through analysis of rate differentials adjusted for different types of dwellings. Assault rates per 1,000 apartment units were exceptionally high in low-status neighbourhoods (20.0 in low SES, 2.3 in high SES). Apart from the pronounced weekend effect discussed above, other calendar effects and those hypothesized for substance abuse could not be substantiated unambiguously. Data inadequacies, particularly with respect to the latter issue, prevented adequate analysis.

Threshold effects

Reviews of the literature of social psychology in the course of our ongoing research revealed an interesting discussion to which we coincidentally found ourselves in a position to contribute. The gist of the argument is found in Anderson and Anderson (1984). Two kinds of behaviour are surmised in response to hot temperatures (as in response to stress in general): fight or flight. The tendency to fight is thought to occur at lower temperatures. Flight occurs above an 'inflection point' beyond which aggression decreases as potential protagonists seek to escape the situation. The Andersons' test of the hypothesis detected a linear relationship between heat and violence, but not a curvilinear one (inverted U shape).

Using a data base extended to cover the period 1 March 1980 through 31 October 1981, with an N of some 10,000, we replicated some aspects of the Andersons' study (Harries and Stadler 1987, in press). Thermal stress was measured by the daily maximum DI and daily maximum ambient temperature. Two regression analyses were performed. In one, an 'overall' model, weekend (operationalized as a dichotomy between weekend and non-weekend), DI, DI squared, maximum temperature and maximum temperature squared were the independent variables. Daily assault count was the dependent variable. DI and DI squared were both significant if entered alone after weekend. With linear DI and maximum temperature controlled, however, their quadratic counterparts were not significant. Disaggregation to the neighbourhood level also failed to show a curvilinear effect after linear variables were entered.

Scatter diagrams of the relationship between daily assault frequencies for the 610 days of the study period and the DI/temperature measures were inspected visually. They suggested a threshold at about 40°C (104°F), and at the corresponding DI, above which daily incidence levels declined. This threshold was independent of the weekend effect. In spite of this apparent threshold, so few days were found at the extreme levels of discomfort that little confidence could be put in this finding. Further study of 6°C (10°F) intervals using analysis of variance and difference-of-means tests also failed to suggest a threshold. Our findings supported those of the Andersons, in so far as weekend was the best predictor of assault frequencies, and thermal effects were linear. Rigorous dissection of this issue, however, would demand neighbourhood-level discomfort data. The failure to detect a curvilinear effect in Dallas using the available data does not prove the non-existence of such a threshold, of course. Such a relationship may exist in other cities in other climatic regimes. Responses to thermal stress may be linked to the concept of normality in different environments. This issue of departures from normal as it relates to the Dallas context has been addressed by Stadler and Harries (1985).

Conclusion

A finding of linear relationship between thermal stress and violence has become commonplace. What research initiatives are now appropriate? Comparative studies are necessary to permit the application of similar measures of violence and thermal discomfort to cities in distinctly different thermal regimes. Such comparisons would provide better understanding of several questions:

(i) Is relative stress important? Put another way, are stress-violence trend lines parallel in different thermal settings? If they are not parallel, is variation random or systematic? If the latter, what factors are at work? (see Kalkstein and Valimont 1986).
(ii) Do apparent linear effects show up in a variety of climatic regimes? The corollary question is: Are linear effects only observed in severe summer regimes?
(iii) Are there lag effects? If so, what are they like, and do they differ across climatic regimes? A lag is understood to mean a short-term delay in the manifestation of stress, and it is socially conditioned. The individual is stressed, but waits until getting off work, for example, before 'blowing up'
(iv) Are there cumulative or additive effects of stress that cause people to 'snap', but in quite different individual time frames? Such cumulative effects are seen as longer term (days, weeks) than lags, and leading to overt reaction relatively late in the period of thermal stress

Such questions should be examined in the context of a concern about violence that has spread across disciplines in recent years. The analysis of violence has moved from being the bailiwick of criminologists, sociologists, and psychiatrists to a broader front embracing various social science fields, including geography. Recently, the public health profession has become involved. Landmarks have included the Surgeon General's Workshop on Violence and Public Health (US Surgeon General 1985), a meeting at the New York Academy of Medicine dedicated to public health perspectives on homicide (Bulletin of the New York Academy of Medicine 1986), and a research conference on violence and public health held in Dallas in 1986.

While the idea that medicine can 'cure' violence is clearly absurd, more modest initiatives incorporating public health perspectives may be constructive. Much, for example, has yet to be learned about the epidemiology of violence, and the concept of violence itself is being broadened to include all kinds of accidental trauma, various types of abuse, as well as the traditionally defined crimes of violence. At present, in the US, no comprehensive standardized data base exists with respect to the emergency room treatment of trauma victims, and it follows from this that detailed standardized data on assaults, homicides, and other acts of violence are lacking. Variations in styles of police reporting forms (e.g. open versus closed-ended) result in important qualitative differences in information recorded for similar incidents occurring in similar circumstances. This lack of standardization then biases subsequent interpretations.

Combination of the perspectives of criminology with those of 'stress' sociologists, 'quality of life' sociologists and geographers, social psychologists, and various public health approaches, yields a rich new synthesis of definition, theory, data, and method. When the geographic matrix is superimposed on this synthesis, fresh insights will be possible.

Acknowledgement

We wish to thank Carl Makres of the Dallas Police Department for supplying the data on which our research has been based. We also appreciate his generous assistance with their interpretation.

References

Anderson, C.A. and Anderson, D.C. (1984) 'Ambient temperatures and violent crime: tests of the linear and curvilinear hypotheses', Journal of Personality and Social Psychology 46, 91-7

Anonymous (1833) Westminster Review 18, 353-6

Baron, R.A. (1972) 'Aggression as a function of ambient temperature and prior anger arousal', Journal of Personality and Social Psychology 21, 183-9

Baron, R.A. and Bell, P.E. (1975) 'Aggression and heat: mediating effect of prior provocation and exposure to an aggressive model', Journal of Personality and Social Psychology 31, 825-32

Baron, R.A. and Ransberger, V.M. (1978) 'Ambient temperature and the occurrence of collective violence:

The "Long, Hot Summer" revisited', Journal of Personality and Social Psychology 36, 351-60

Boyanowski, E.O., Calvert, J., Young, J. and Brideau, L. (1981) 'Towards a thermoregulatory model of violence', Journal of Environmental Systems 1, 81-7

Braucht, G.N., Loya, F. and Jamieson, K.J. (1980) 'Victims of violent death: a critical review', Psychological Bulletin 87, 309-33

Bulletin of the New York Academy of Medicine (1986) 'Homicide: the public health perspective', 62

Bureau of Justice Statistics (1980) Crime and Seasonality, Washington DC: US Government Printing Office

Bureau of Justice Statistics (1983) Report to the Nation on Crime and Justice, Washington DC: US Government Printing Office

Calvert-Boyanowski, J., Boyanowski, E.O., Atkinson, M., Goduto, D. and Reeves, J. (1976) 'Patterns of passion: temperature and human emotions', in Krebs, D. (ed.), Readings in Social Psychology: Contemporary Perspectives, London: Harper & Row, 96-8

Carlsmith, J.M., and Anderson, C.A., (1979) 'Ambient temperature and the occurrence of collective violence: a new analysis', Journal of Personality and Social Psychology 37, 337-44

Cohen, J. (1941) 'The geography of crime', Annals of the American Academy of Political and Social Science 217, 29-37

Cohen, S. (1980) Folk Devils and Moral Panics, London: MacGibbon and Kee

Cohen, L.E. and Felson, M. (1979) 'Social change and crime rate trends: a routine activities approach', American Sociological Review 44, 588-608

Dexter, E.G. (1904), Weather Influences: An Empirical Study of the Mental and Physiological Effects of Definite Meteorological Conditions, New York: Macmillan

Federal Bureau of Investigation (1986), Uniform Crime Reports, US Government Printing Office, Washington DC. (These reports are published annually).

Gabor, T. (1986) The Prediction of Criminal Behavior: Statistical Approaches, Toronto: University of Toronto Press

Glacken, C.J. (1967) Traces on the Rhodian Shore, Berkeley: University of California Press

Harries, K.D. (1976) 'A crime-based analysis and classification of 729 American cities', Social Indicators

Research 2, 467-87

Harries, K.D. (1985) 'The historical geography of homicide in the US', Geoforum 16, 73-83

Harries, K.D. (1986) 'Serious assault in Dallas, Texas: the roles of spatial and temporal factors', Journal of Justice Issues 1, 31-45

Harries, K.D. and Stadler, S.J. (1983) 'Determinism revisited: assault and heat stress in Dallas, 1980', Environment and Behavior 15, 235-56

Harries, K.D. and Stadler, S.J. (1987) 'Heat and violence: new findings from Dallas field data, 1980-81' Journal of Applied Social Psychology in press

Harries, K.D., Stadler, S.J. and Zdorkowski, R.T., (1984) 'Seasonality and assault: explorations in inter-neighbourhood variation, Dallas, 1980', Annals, Association of American Geographers 74, 590-604

Holt-Jensen, A. (1980) Geography: Its History and Concepts, London: Harper & Row

Huntington, E. (1915) Civilization and Climate, New Haven: Yale University Press

Huntington, E. (1945) Mainsprings of Civilization, New York: John Wiley & Sons

Huntington, E. (1907) The Pulse of Asia, New York: Houghton-Mifflin Company

James, P.E. and Martin, G.J. (1985) All Possible Worlds: A History of Geographical Ideas, 2nd edn, New York: John Wiley & Sons

Kalkstein, L.S- and Valimont, K.M., (1986) An evaluation of summer discomfort in the United States using a relative climatological index, Bulletin of the American Meteorological Society 67, 842-8

Koller, A.J. (1937) The Abbé du Bos - his advocacy of the theory of climate, Champaign, Ill.: The Garrad Press

Landsberg, H.E. (1969) Weather and Health: An Introduction to Biometeorology, New York: Doubleday Anchor Books

Lee, D.H.K. (1954) 'Physiological climatology' in James, P.E., and Jones, C.F. (eds), American Geography: Inventory and Prospect, Association of American Geographers, Syracuse, New York, 470-83

Lewis, L.T. and Alford, J.J., (1975) 'The influence of season on assault', The Professional Geographer 28, 214-17

Linsky, A.S. and Strauss, M.A. (1986) Social Stress in the United States, Dover, Mass.: Auburn House

Loftin, C. and Parker, R.N. (1985) 'An error-in-variable model of the effect of poverty on urban homicide rates'

Criminology 232, 269-87
Martin, G.J. (1973) Ellsworth Huntington: his Life and Thought, Hamden, CT: Archon
Messner, S.F. and Tardiff, K. (1985) 'The social ecology of urban homicide: an application of the "routine activities" approach' Criminology 23, 241-67
Mills, C.A. (1942) Climate Makes the Man, New York: Harper
Oliver, J.E. (1973) Climate and Man's Environment New York: John Wiley & Sons
Oliver, J.E. and Siddiqi, A.K. (1987) 'Climatic determinism' in Oliver, J.E. and Fairbridge, R. (eds), The Encyclopedia of Climatology, Encyclopedia of Earth Sciences Series, XI, New York: Van Nostrand Reinhold Company
Persinger, M.A. (1980) The Weather Matrix and Human Behavior, New York: Praeger
Ratzel, F. (1882-1891) Anthropogeographie, 2 vols, Stuttgart: Englehorn
Rotton, J. (1986) 'Determinism redux: climate and cultural correlates of violence', Environment and Behavior 18, 346-68
Rotton, J. and Kelly, I.W. (1985) 'Much ado about the full moon: a meta-analysis of lunar-lunacy research', Psychological Bulletin 97, 286-306
Sampson, R.J. (1986) 'Crime in cities: the effects of formal and informal social control', in Reiss, A.J. and Tonry, M. (eds), Communities and Crime, Chicago: University of Chicago Press, 271-311
Smith, D.M. (1973) The Geography of Social Well-Being New York: McGraw-Hill
Spate, O.H.K. (1952) 'Toynbee and Huntington: a study in determinism', Geographical Journal 118, 406-28
Stadler, S.J. and Harries, K.D. (1985) 'Assault and deviations from mean temperatures: an inter-year comparison in Dallas', Papers and Proceedings of Applied Geography Conferences 8, 207-15
Stoddart, D.R. (1972) 'Darwin's impact upon geography' in Stoddart, D.R. (ed.) The Conceptual Revolution in Geography, Totowa, New Jersey: Rowan & Littlefield, 52-76
Sutherland, E.H. and Cressey, D.R., (1978) Criminology, 10th edn, Philadelphia, PA: Lippincott
Taylor, G. (1955) Australia, London: Methuen
Toynbee, A.J. (1955) A Study of History 10 vols, Oxford:

Oxford University Press

Tromp, S.W. (1980) Biometeorology: the impact of Weather and Climate on Humans and Their Environment, London: Heyden & Son Ltd.

US Surgeon General (1985) Surgeon General's Workshop on Violence and Public Health, Report, US Public Health Service, Washington DC

Chapter Four

MICRO-ENVIRONMENTS OF VIOLENCE

R.N. Davidson

Throwing a punch is an act of violence. Given that it connects, and even perhaps if it does not, it has consequences. And so it should. The problem is that the consequences are not uniform nor consistent, may indeed vary dramatically with the circumstances that surround the act. Consider the following scenarios: a punch thrown at a meeting convened for some other purpose, classroom, party caucus, wedding; a punch thrown at you in your street; a punch thrown in a padded ring occupied by two men wearing gloves. The act may be the same, but the significance of the act for all concerned will be very different.

The punch at the private gathering will probably result in the intervention of someone in authority, teacher, party leader, best man, in an attempt to mediate, resolve the dispute and return to the business in hand. The involvement of the police is a last resort if matters get out of hand. If injury is not severe, there may be few consequences for either victim or assailant. If, however, it is the teacher or party leader who throws the punch, a very different light may be cast on the act. Violence in most situations is very volatile but more so in private between people who are connected in some way. Being knocked to the ground in your own street is more straightforward especially if the assailant is a stranger, an outsider. You, or anyone else who witnesses the incident, will probably involve the police directly - the more serious the blow, the more quickly the call will be made. If it is not your street, matters become less simple and if in addition there are no witnesses, you may be much more reluctant to involve authority. Will you be believed? Can anything be done? Perhaps it would be better to slink off home and try to forget about it. The emphasis on you, here, is quite deliberate to reinforce the fact that personal feelings are highly involved and interact with the environment. Safety in your street is a matter of greater significance to you than safety in someone else's street. It is certainly not threatened by the punches thrown by the boxers in the ring. Here there is no clear-cut distinction between assailant and victim, and nobody is calling for the involvement of the criminal justice system. Indeed, the punch will be applauded if it connects and there might be cries for more, at least among a section of the audience.

These illustrations serve to illuminate the inextricable involvement of geography in understanding patterns of violence. There is no way that acts which constitute violent behaviour can be separated from the settings in which they occur. McClintock (1963) recognized this a quarter of a century ago when he developed a situational typology of violence in order to highlight the very different circumstances that surround incidents. McClintock's study was based on police reports in London and the three main situational categories he adduced were domestic and neighbourhood disputes (about 31 per cent), fights in pubs, etc. (20 per cent) and violence in public places (30 per cent). While we may now criticize McClintock's typology as a strait-jacket which submerges alternative qualities of places, such as privacy and familiarity, we must acknowledge the debt owed to him for abandoning the strictly legal categories of violence.

The situational theme was picked up and extended by other writers such as Lambert (1970) and Normandeau (1969) dealing with minor disputes and robberies respectively. By the 1980s, these and other piecemeal efforts on both sides of the Atlantic had merged into a school of environmental criminology (Davidson 1981; Brantingham and Brantingham 1981) which sought to codify the multifarious relationships between crime and its setting at various spatial scales. At the same time, interest has grown in situational crime prevention (Clarke 1980; Heal and Laycock 1986) in which the understanding of crime in its place is seen to underpin opportunities for reducing crime by managing the environment. Studies such as those of Poyner (1981) and Ramsey (1982) on city-centre crime have an explicit crime-prevention orientation. There is a need, however, to return to McClintock's roots and examine the basic patterns of violence at a high level of spatial resolution in order to reaffirm the detailed parameters of the relationships between violence and its setting. The objective of this chapter is an evaluation of the resources of the second (1984) British Crime Survey in this task, looking particularly at assault. Other offences with an element of violence, such as aggravated burglary, robbery or sexual offences are excluded from this analysis.

Any study of violence is bedevilled by two problems which can never be properly resolved. One is that of obtaining adequate information. Police reports are incomplete: less than half of the incidents of serious violence may be reported to the police, for less serious violence the proportion may fall below one third (Hough and Mayhew 1985). Victim surveys may be more complete, but still suffer from deficiencies in particular areas, for example domestic violence or sexual assault (Jones et al. 1986). There is no easy remedy to the unrevealed dark figure of violence, not least because even the admission of victimization may be a source of stigmatization among the young men who are most at risk. The particular problems of using the British Crime Survey are discussed below.

The other problem is interconnected but even more difficult to handle - the categorization of victims and offenders. The law demands this, provides for the rigorous separation of parties in court, but the reality is not so clear-cut. The act of violence that lands someone in court is likely to be the culmination of an interaction between individuals involving other behaviour just prior, even perhaps well before, the sanctionable act. Striking the first blow is not the sole nor necessary criterion but in evidential terms it is the most easily sustainable and therefore critical. Striking the first blow may be a response to other provocations, connected in neither time nor space to the present circumstances, for example the culmination of racial harassment. This problem is made more difficult by the degree of intermingling of offenders and victims evident from most studies of violence - the similarity of age, lifestyle and social position (Hindelang et al. 1978; Gottfredson 1981). The British Crime Survey data show that the likelihood of assault victimization is seven times as high among those who admit offending (Gottfredson 1984). Care must be taken to examine, as far as is possible, all aspects of the circumstances and not concentrate simply on victims.

Environmentally relevant criteria in violence

Any investigation of the setting of assault needs to accommodate a number of different perspectives on the relevant criteria. From the victim's point of view, the most important are the functional properties of the place where the incident occurs - these indicate the reason for the victim's presence. Such criteria may be loosely termed the behavioural environment. Monahan and Klassen (1982) amongst others have emphasized three prime behavioural settings each with particular correlates with violence; the family environment, the peer environment and the job environment.

Settings also have physical properties which may be relevant. The amount of space, degree of crowding, lighting, and other design features that influence what happens (Poyner 1983). Likewise there are perceptual elements of places such as feelings of safety or fear, familiarity, and privacy. Stokols (1979) suggests that setting is important in facilitating or restraining psychological needs and goals. The relevance of environmental factors to both attacker and victim must be emphasized.

The wider environment must also be considered. Assault has a social context as well as an environmental setting. Factors such as alcohol, pals, and the relationship between attacker and victim prior to the incident may precipitate an incident. Beyond this, both parties may bring with them to the setting personal characteristics which may predispose them, actively or passively, towards violent behaviour. Gender, age, social position, and psychological traits are all relevant in differentiating risks of exposure to violence (Gottfredson 1984). They are also relevant in differentiating the consequences of the event (Davidson 1986).

The purpose of this analysis is to examine the contribution of the wider neighbourhood to exposure to violence and to assess the influence of the immediate setting on the outcome of and reactions to the event.

The British Crime Survey as data source

The data used in this analysis is drawn from the second (1984) British Crime Survey (BCS), a national survey of the victims of crime. The BCS used a stratified random sample of some 11,000 households (NOP, 1985) which yielded details of some 448 victims of assault or attempted assault during the 14-month period of the survey. In the present analysis, the data have been treated in two ways. For the neighbourhood patterns, the base is victims: for the analysis of settings the base is incidents, that is including a weighting for repeated or series assaults. In both cases, the data are weighted for bias in the sample fractions but not adjusted to annual rates in order to preserve as large a sample as possible. Significance tests, where applied, are based on the chi-square statistic. An arbitrary minimum of 10 cases is used where rates are calculated, though frequencies of less than this may be significantly lower than expected.

The BCS presents a very large number of variables for analysis and there are many ways each may be categorized. The classifications have been kept very simple because the number of assaults is not sufficiently large to allow more detailed examination. Not all the variables available are presented here. Some do not add to the analysis, others such as victim's and offender's race were too attenuated to permit viable conclusions. Not all environmentally relevant information was recorded by the BCS, particularly the reasons for being there and perceptions about the place.

Despite its thoroughness, the BCS is bound to provide an underestimate of crime levels in general (Hough and Mayhew 1985) and of violence in particular. There may be bias arising from differences in ability to complete complex questionnaires (Hough and Sheehy 1986). There may be widespread differences in the interpretation of what happened: what is acceptable to male youth on a Saturday night, and not therefore recorded by the survey, may be a traumatizing experience to an old lady. With certain categories of violence the problem is magnified: accessing domestic violence by a household interview is not likely to be very successful. As Walmsley (1986) suggests, there is no clear body of knowledge on domestic violence. The fact remains that the BCS is a more comprehensive source than its alternatives. Its limitations, nevertheless, need to be made explicit and worked within.

Neighbourhood variations in assault

The uneven geographical distribution of assault is illustrated

Table 4.1 Likelihood of assault victimization: area and victims' gender and age differences

Neighbourhood type (Acorn group)	Assault victims per 10,000 persons 16+	Relative differences in victimization rates (national average = 100)			
		Gender		Age	
		Male	Female	Under 30	30+
Low-risk areas					
A Agricultural areas	225	109	<	218	<
K Better-off retirement areas	253	123	<	<	<
E Better-off council estates	309	134	<	197	34
Medium-risk areas					
B Modern family housing higher income	364	171+	-	262	36
F Less well-off council estates	379	135-	58+	280	44
J Affluent suburban housing	400	151	50	304	51
C Older housing of intermediate status	430	177	37	278	54

Table 4.1 continued

High-risk areas					
D Poor-quality older terraced housing	515	171	86	250	77
I High status non-family areas	527	232	<	294	70
H Multi-racial areas	564	210	<	323	<
G Poorest council estates	584	154-	135+	334	<
National Average	404	161	43	266	46

Notes

< too few victims (<10) to generate reliable rates.

\+ significantly more victims (at p=0.05 level) than expected from national differences.

\- significantly fewer victims (at p=0.05 level) than expected from national differences.

Source

British Crime Survey 1984, data weighted but not adjusted for 14-month survey period.

Table 4.2 Likelihood of assault victimization: area and victim status differences

Neighbourhood type (Acorn group)	Relative differences in victimization rates (national average = 100)								
	H'hold Head's Occupation		Employment status			Age education ended		Housing	
	Non-manual	Manual	Employed	Unem-ployed	In-active	<17	17+	Own-occup.	Other tenure
Low-risk areas									
A. Agricultural areas	<	<	93	<	<	<	<	<	<
K. Better-off retirement areas	<	<	108	<	<	<	<	<	<
E. Better-off council estates	<-	92+	91	<	<	60	167	77	72
Medium-risk areas									
B. Modern family housing higher income	93+	89-	120+	<	<	90	90	73	185
F. Less well-off council estates	<	98	131	<	<	81	<	63-	107+
J. Affluent suburban housing	94+	105-	123	<	60+	84-	123+	102+	85-
C. Older housing of intermediate status	83	121	152	<	26	105	116	114+	86-

Table 4.2 continued

High-risk areas									
D. Poor-quality older terraced housing	<	135	177	<	<	118	<	93	203
I. High status non-family areas	115+	<	181	<	<	128-	134+	79-	194+
H. Multi-racial areas	<	131	165	<	<	138	<	<	<
G. Poorest council estates	<	157	<	<	141+	130	<	<	152+
National average	87	107	131	205	42	92	123	98	119

Notes

< too few victims (<10) to generate reliable rates.

+ significantly more victims (at p = 0.05 level) than expected from national differences.

- significantly fewer victims (at p = 0.05 level) than expected from national differences.

Source

British Crime Survey 1984, data weighted but not adjusted for 14-month survey period.

by Tables 4.1 and 4.2 in which differences in the likelihood of becoming a victim are mapped using the ACORN classification of residential neighbourhoods. Low-risk neighbourhoods include agricultural areas, better-off retirement areas and better-off council estates. As a whole such areas have half to three-quarters the risk of assault compared to the national average. At around the national average, medium-risk areas include modern family housing of higher income, less well-off council estates, affluent suburban areas and older housing of intermediate status. High-risk neighbourhoods, with up to one-and-a-half times the national rate, include poor-quality older terraced housing, high-status non-family areas, multi-racial neighbourhoods and the poorest council estates. Compared to the risks of burglary (Hough and Mayhew, 1985, p. 37), assault differentials are rather lower - about half - indicating a wider spread of assault victimization. The ranking of neighbourhood types is, however, broadly similar between assault and burglary with one or two notable exceptions. Risk of assault in older housing of intermediate status is comparatively much higher than the risk of burglary. Better-off council estates on the other hand are much safer for assault than for burglary.

These broad differences conceal significant variations between neighbourhoods in the type of victim likely to be attacked. Gender is an important factor in assault victimization, with males approximately four times more at risk than females. This differential, however, is not reflected across all neighbourhood types. Female victimization is particularly high in areas with council estates - indeed in the poorest council estates there is virtual parity between male and female risks and the female risk here exceeds the male risk among low-risk neighbourhoods. These gender differentials may help to explain some of the contradictions between the BCS and local crime surveys. For example the Islington Crime Survey (Jones et al. 1986), which reports an overall assault rate higher for females than for males, was conducted in an area with a high proportion of public housing.

Age is also an important discriminator of assault victimization, with younger victims more than five times more at risk than older. In this case, however, there are no significant variations in the rates between the neighbourhoods. Patterns of occupational differentiation are clearly demarcated (Table 4.2). Victims from households with a non-manual head are over-represented in higher-status neighbourhoods. Conversely, though not so dramatically, it is lower-status households which suffer assault victimization in lower-status neighbourhoods. What is being demonstrated here quite clearly are separate effects for neighbourhoods and for households. In terms of victim's employment status, differentiation is much less marked. Unemployed people suffer twice the risk of assault on average but lack of numbers in this category precludes area by area breakdown. Victims in employment are significantly over-represented only in modern high-income areas, whereas risks for the economically inactive (housewives, students, etc.) are high in two extremely contrasting social areas -- affluent suburbs and the poorest council estates.

Education and housing differences strengthen the pattern of social polarization of assault risks. Victims who pursued their education beyond the minimum age and those who live in owner-occupied housing can suffer enhanced risks if they reside in a better-off neighbourhood, particularly the affluent suburb. The trend, however, is by no means straightforward: owner-occupiers in high-status non-family areas have reduced risks whereas those from older housing of intermediate status have higher than expected risks. It is easy to conclude from these data that neighbourhood is relevant to assault victimization, but the effects are complex and not amenable to more intensive anlaysis as numbers become too small beyond the simple categorizations employed here.

Neighbourhood profiles of assault incidents

The nature and circumstances of the assaults suffered vary from neighbourhood to neighbourhood. The profiles presented here summarize a wide-ranging set of characteristics of the incident - the setting of the assault, who was involved, the consequences for the victim in terms of injury and other problems, and the reaction of victim and police involvement. Again data attenuation is a problem, with many categories having too few incidents in some areas for analysis. The following comments are based on significant neighbourhood deviations from national trends (more or fewer incidents than might be expected if the national pattern were applied to each neighbourhood type).

Absence of comment does not necessarily imply absence of differentiation: it can also mean that the data were too sparse to infer any trend.

Data attenuation is particularly severe in low-risk agricultural and better-off retirement areas where there are simply too few incidents to attempt a breakdown. With better-off council estates there is a slight trend for assaults to take place indoors, in private, and to be connected with the victim's workplace. No injury is the common result.

Assaults on victims from modern family housing areas of higher income exhibit a strong tendency to take place outdoors, in the street and to be connected with pubs, clubs, and other places of entertainment. Assailants are often previously acquainted with the victim, but not a relative. Victims claim little provocation was offered and there were no emotional problems afterwards. All this stands in contrast to the pattern of less well-off council estates where a high proportion of assaults take place in the victim's home and where assailants are often related to their victims. Injury is common, and many of the victims report suffering emotional problems afterwards. The trend for affluent suburban areas is nearly as strong but again different, being associated with both workplace and entertainment (and both indoors and on the street). Normally only one incident is reported, by a stranger. No provocation is offered, no injury is common, as is a lack of emotional problems afterwards. This pattern is repeated for assaults on victims from older housing of intermediate status, though more of the incidents resulted in minor injury and, curiously, many of the victims offered some provocation.

Contrasts among the high risk areas are equally distinctive. Victims from poor-quality older terraced housing areas tend to suffer more from the domestic type of violence, with assaults by a relative being two-and-a-half times more common. A similar pattern applies to the poorest council estates with more than twice as many assaults taking place in the home as might be expected. However, unlike the other neighbourhood types where domestic violence predominates, in these poorest estates attacks in the street are also more common. High-status non-family neighbourhoods are not associated with a strongly differentiated pattern of violence showing only weak connections with places of entertainment and for incidents to take place in the street.

The final neighbourhood type, multi-racial areas, shows a unique profile of violence from the BCS. This is the only residential type where incidents tend to be connected with places other than home, workplace or entertainment (though they do take place predominantly in the street). More significantly, multi-racial districts suffer an unusually large number of assaults where there is more than one offender and in which incidents are not isolated but are repeated. Regrettably, such areas cover such a small fraction of the population (3.9 per cent) at national level within the ACORN system that, even with a high incidence of assault, there are simply too few incidents being reported in the BCS to provide further illumination.

In general, these neighbourhood profiles suggest an emphasis on domestic violence on council estates and in inner-city areas. Among more affluent areas, pubs and the workplace serve as more significant focal points for violence.

Micro-locational settings of assault

The British Crime Survey provides a detailed classification of the places where assaults are perpetrated. Violence is concentrated (Table 4.3) in or near the three locations at which people spend most time in normal activities - home, work and pubs, discos and other places of entertainment. The remaining one-third of incidents occur in a wide variety of other locations, from beaches to churches, including a small but significant number where the victim's recollection was poor or uncertain. This analysis will focus on home, work, and pub as contrasting behavioural settings - other locations omitted to sharpen differences. An alternative perspective is provided by the physical setting in which the basic division is between indoor and outside locations with each being further subdivided according to whether the incident occurred in public or private space. Only vague or unknown locations were omitted from this secondary classification. Of known locations 42 per cent were outside, 34 per cent in the street. Fifty-eight per cent of assaults took place indoors, of which the majority, 37 per cent, were in a private place (home or work). In all 55 per cent of incidents took place in public settings.

Table 4.3 Assault settings

Home		
inside own home	65	
outside own home	40	
in/near relative/friend's home	32	
		137
Work		
inside	90	
car park, street near	24	
		114
Pub, disco, other entertainment		
inside	82	
car park, street near	56	
		138
Other street, park, car park, waste ground		
		74
Other premises (factory, shop, school, hospital, etc.)		
in or near		37
Public transport		
in or near		14
Elsewhere, vague, not known		
		85
All incidents		599

Source
British Crime Survey 1984, data weighted.

Table 4.4 enumerates the degree of injury suffered in different settings. Violence connected with homes results in more minor injury than the other behavioural settings. Serious or multiple injury is less common, as is no injury. This setting is clearly the most problematic for surveys such as the BCS which are based on household interviews. Surveys based on assaults reported to the police show a higher proportion of serious injury in domestic incidents which can be accounted for by two rather different reasons. First, less serious domestic violence does not get reported to the

Table 4.4 Assault setting and injury caused

	Injury caused (percentages)			
	Multiple/ serious	Minor	None	Total
Behavioural setting:				
In or near victim's home (N=105)	15	46	39	100
In or near victim's place of work (N=114)	18	26	56	100
In or near pub, disco, etc (N=138)	26	38	36	100
Physical setting:				
In the street (N=169)	36	34	30	100
Other outside place (N=43)	14	33	53	100
Indoors, public place (N=107)	23	35	42	100
Indoors, private place (N=184)	13	35	52	100
All assaults (N=599)	22	34	43	100

Source
British Crime Survey 1984, data weighted.

police (Davidson 1982) or if it is it may subsequently be 'no-crimed' (cf. Bottomley and Coleman 1981). Second, victimization survey respondents may be afraid to report more serious incidents out of fear.

Assault related to the workplace involves no injury significantly more frequently than elsewhere. The fact that the proportions of both serious and minor injury are reduced suggests that this is a threshold effect: victims are more likely to regard a minor fracas as assault than they might in other situations. The injury profile for pub and disco settings tends to confirm this suggestion that tolerance of violence varies with setting. Here both minor and serious injury are increased in proportion. What is tolerated in the pub may not be tolerated at work. Victims are significantly less likely to report incidents to the police in the pub or disco context than at home or work, though about the same proportion of assaults in all three settings come to the attention of the police.

These observations are sharpened by the effects of physical setting on injury. The significant difference here is between private and public places regardless of whether they are indoors or outside. In all cases the variation is between the 'serious' and 'no injury' categories. Attacks in public places both indoors and particularly outside have a higher injury content: attacks in private are less likely to result in injury. However, the tolerance hypothesis for these differences is not so easy to sustain. If it were true, then lower tolerance in private settings would produce an enlarged no injury category and thereby depress the proportions of both injury groups. This is not the case, so other factors must be in play, for example, the influence of privacy on the interaction itself - on the way in which both assailant and victim relate to their immediate surroundings. For victims privacy reduces opportunities for help, whether it is others intervening or more positively being brought in as a means of reducing tension. For assailant, privacy increases security, reduces fear of being caught. The combination of these influences places the assailant-in-private in a more powerful position in his/her aggression. There is therefore less need for the infliction of pain to achieve ends, threat is sufficient. In public places assailants need to overcome their own fears as well as the victim. In surmounting this threshold more people get seriously hurt. Such a threshold of force hypothesis is by no means sufficient explanation for the patterns. For one thing it assumes, to use Block's (1977) terminology, instrumental rather than affective violence, that is, it is a means to an end. We do not know enough about assailants' motives from the BCS to be able to ascertain this. Secondly, the influence of privacy is counteracted by that of familiarity of both victim and assailant with their surroundings (Davidson 1982). The inferiority of victims in private can be offset by familiarity especially if the assailant is less familiar (and knows it). The threshold of violence may thus be lowered if the victim is in his/her own home or some other place where they have a right to be and where the assailant does not. Regrettably the BCS data do not allow the familiarity aspects of the setting to be elaborated. It is, however, abundantly clear that the setting is an important determinant of the outcome of violence.

Victim risks and assault settings

The differential exposure to risks of violence have been related to a number of personal variables (Hough and Sheehy 1986; Gottfredson 1984; Hindelang et al. 1978). Among these, gender, age, status, and lifestyle have been seen as particularly relevant. We need now to connect these influences to those of setting so that the lessons to be derived from environmental cues may be delivered to the potential victims most at risk. Tables 4.5, 4.6, and 4.7 set out the incidence of assault in the various settings for different types of victim. All the rates are presented per 10,000 persons aged 16+ in the relevant category in order to facilitate comparisons. The baseline rate of assault victimizations is 540 assaults per 10,000 persons 16+ over the 14-month survey period.

Table 4.5 shows age and gender differences among the settings. As a whole, males have more than three times the risk of assault as females, and young people (under 30 years) over five times the risk of older people. Age differences are much more exaggerated among females, with young women having eight times the risk of older. Indeed when specific settings are examined, young women are exposed to greater risk of violence in the home than older men anywhere and young men in most settings. Bearing in mind the underestimation likely in the BCS of this very setting, these rates emphasize in an unequivocal way the problem of domestic violence. Moreover, unless it can be assumed that the underestimation is confined to older women - a very dubious assumption - the data highlight the specific exposure of young women to assault in the home.

For younger men the most dangerous setting is the pub or disco, with about 35 per cent of assaults taking place in or near them. Violence related to workplaces is notable for the fact that risks are much less significantly reduced with age than in other behavioural settings. Physical settings reveal a further dimension to the gender dichotomy - males tend to be assaulted in the street, females indoors in private. Again this is a pattern that is not apparent among older victims where private, indoor settings are commonest for both men and women.

Occupation of the victim's head of household and victim's own employment status provide further illustration of differential risks (Table 4.6). Victims from non-manual households tend to be assaulted in or near their workplace

Table 4.5 Incidence of assault: setting and victim's age and gender

Assaults per 10,000 persons 16+	Aged under 30			Aged 30 or over			All victims
	Male	Female	Total	Male	Female	Total	
All incidents (N=599)	2,080	801	1,448	419	102	251	540
	All males 844			All females 262			
Behavioural setting							
In or near victim's home	125	393	258	56	31	42	95
In or near victim's place of work	236	(38)	138	168	27	93	104
In or near pub, disco, etc	715	136	430	48	(4)	25	123
Physical setting							
In the street	708	159	437	96	31	62	152
Other outside place	125	(23)	75	43	(13)	27	39
Indoors, public place	560	76	321	48	(4)	25	96
Indoors, private place	324	461	396	152	43	94	166

Notes
Figures in brackets indicate too few (<10) incidents to generate reliable rates.
Column figures may not add due to exclusion of other settings or missing data.

Source
British Crime Survey 1984, data weighted but not adjusted for 14-month survey period.

Table 4.6 Incidence of assault: setting and victim's occupation and employment status

Assaults per 10,000 persons 16+	H'hld Head's Occupation		Employment status		
	Non-manual	Manual	Employed	Unemployed	Inactive
All incidents (N=599)	537	547	704	1,117	235
Behavioural setting					
In or near victim's home	66	118	84	275	82
In or near victim's place of work	171	62	181	(16)	(9)
In or near pub, disco, etc	86	148	165	210	50
Physical setting					
In the street	121	184	165	518	87
Other outside place	41	41	56	(81)	(9)
Indoors, public place	73	105	139	162	30
Indoors, private place	184	152	232	178	75

Notes

Figures in brackets indicate too few (<10) incidents to generate reliable rates.
Column figures may not add due to exclusion of other settings or missing data.

Source

British Crime Survey 1984, data weighted but not adjusted for 14-month survey period.

Table 4.7 Incidence of assault: setting and victim's housing status and income

Assaults per 10,000 persons 16+	Housing status		Income		
	Own-occupied	Other tenure	£5,000	£5,000–£9,999	£10,000+
All incidents (N=599)	493	630	360	374	521
Behavioural setting					
In or near victim's home	36	206	212	90	(36)
In or near victim's place of work	127	58	(7)	129	168
In or near pub, disco etc	123	121	43	70	112
Physical setting					
In the street	141	177	133	90	120
Other outside place	38	40	(22)	51	(28)
Indoors, public place	88	113	47	(35)	92
Indoors, private place	133	229	158	187	196

Notes
Figures in brackets indicate too few (<10) incidents to generate reliable rates.
Column figures may not add due to exclusion of other settings or missing data.

Source
British Crime Survey 1984, data weighted but not adjusted for 14-month survey period.

and indoors in private compared to victims with manual household heads where pub or disco risks are greater (though not, curiously, as high as workplace risks for the non-manual) and the street is the commonest physical setting. Employed people have the greatest risk of violence related to work or to pub or disco. For the unemployed and economically inactive, risks focus on the home and in both cases the street is the dominant physical setting. It seems clear that being employed is a further basic differentiator of risks among the settings. The unemployed may suffer enhanced risks as a group but the pattern is similar to that of the economically inactive. Housing status and income (Table 4.7) confirm the distinction between work and pub on one hand and the home on the other. Victims from owner-occupied housing are more exposed to violence at work or at pub and victims from rented accommodation have greater risks in the home. It is interesting to note that risks of assault increase with household income, though not substantially, but this conceals two diametrically opposed patterns. Low-income victims face the greatest risks in the home, high-income victims at work. Physical settings, however, show no consistent relationship to income.

The outcome of violent interactions in terms of injury risks is illustrated in Table 4.8 by victim's age and gender. While males have higher risks in all categories, women seem to suffer relatively higher risks of minor injury in both age groups. These overall patterns simply reflect the differential involvement of women in various settings. This point is reinforced by information, also given in Table 4.8, on victim-offender relationships. This highlights the gross disparity between younger men and women victims in terms of who perpetrates the attack. Well over half the assaults on young women are by relatives, including live-in partners. For men, most assailants are strangers. The incidence of assault by relatives for young women exceeds that of older men by all types of assailant. The fact that attacks by relatives on older women are very rare in the BCS needs careful evaluation. If it is due to reluctance among respondents to admit to this kind of victimization and this reluctance is not age-specific, then the rate for younger women will need considerable upward revision to a realistic level.

Table 4.8 Incidence of assault: victim's age and gender, injury and victim-offender relationship

Assaults per 10,000 persons 16+	Aged under 30			Aged 30 or over			All
	Male	Female	Total	Male	Female	Total	victims
All incidents (N=599)	2,080	801	1,448	419	102	251	540
		All males 844			All females 262		
Injury caused							
Multiple/serious	531	151	343	91	(11)	49	119
Minor	730	355	545	109	47	75	189
None	819	295	560	218	47	127	231
Victim/offender relationship							
Relative	73	499	287	(8)	(11)	(10)	77
Acquaintance	870	151	515	160	49	101	201
Stranger	1,136	151	646	249	43	140	262

Notes

Figures in brackets indicate too few (<10) incidents to generate reliable rates.
Column figures may not add due to missing data or rounding errors.

Source

British Crime Survey, 1984, data weighted but not adjusted for 14-month survey period.

Profiles of assault in different settings

Just as different sorts of victims are exposed to risk of violence in different settings, so the circumstances and responses to incidents may vary. Rather than present a large and complex array of data on the fifteen measures examined, the strategy is again adopted to profile the settings in terms of the salient categories which emerge from the cross-tabulations.

Of the behavioural settings, assaults in or near the victim's home are characterized by a greater daytime frequency but are spread throughout the week. Most involve one assailant and are part of a series which relate to the fact that the attacker was a relative in about three out of five incidents, and had the right to be there in about the same proportion. There is no evidence that domestic incidents are less likely to come to the attention of the police, but if they do so, the victim is more likely to be the informant than in other settings. As we have seen, domestic victims are more likely to suffer injury, though serious injury is less common. However, more victims in domestic situations report both practical and emotional problems after the attack - about 70 per cent in the latter. Assaults in or near the victim's workplace are also more frequent in daytime hours and, as would be expected, during weekdays. Again incidents that are part of a series are more common, but in this setting the attacker is more likely to be an acquaintance (about half, the remainder being strangers). There is no tendency for the police to be more frequently involved but if they are the victim was more likely to inform them. The injury content of the workplace assaults is low - more than half the victims report no injury - and a higher than expected proportion report no emotional problems afterwards. As might be anticipated, incidents in or near pub, disco, or other place of entertainment provide many contrasts. Attacks are concentrated in the evening and night, and at the weekend. More than one attacker is more common though still only in 45 per cent of cases. The majority of attacks are perpetrated by strangers with acquaintances in a substantial minority. Police involvement is no more likely than in other behavioural settings, but if they are, someone other than the victim called them or they happened to be present. Despite the apparently much higher content of injury in this setting, 80 per cent of victims report no subsequent emotional problems.

The physical settings reinforce some of the patterns highlighted by the behavioural contexts while providing contrasts of their own. Attacks in the street occur much more frequently during the evening and night. A majority are one-off incidents, are perpetrated by strangers, and involve more than one assailant. Such attacks more often come to the attention of the police, perhaps because of their high injury content. This is the only category of assault to produce significant practical problems afterwards, but curiously a lower than expected ratio of emotional problems. Other outside places are rather rare for detailed analysis but do show a tendency towards day-time attacks, one-off incidents, and the involvement of acquaintances rather than strangers. Attacks indoors in a public place tend to occur during evening or night-time hours, be single offences and by either acquaintances or strangers. The police are less likely to be involved and there are fewer emotional problems for victims afterwards. Indoors in a private place produces very strong patterns concentrating on single offender, series attacks, involving relatives (86 per cent of relatives attacked in this setting), with emotional problems after (50 per cent of all victims suffering subsequent emotional problems were attacked in this setting). The police become involved rather less often in indoor, private assaults.

Physical settings provide some consistent differentiation of assaults. Evening and night-time attacks tend to occur in public places, as do stranger assaults. Weekday attacks are more common outside, weekend attacks indoors. The police are more likely to become involved with outdoor as opposed to indoor incidents, but if they are informed, the victim is likely to do it in private settings, other persons or the police being there in public places. Equally important are the criteria which consistently show no differentiation across the settings. Provocation of attackers and the use of a weapon are especially notable, particularly the latter which is often taken as a powerful influence on the outcome of violence (Walmsley 1986; Zimring and Zuehl 1986). Medical attention for injuries was likewise not differentially sought among the settings.

A situational approach to the avoidance and prevention of violence

The BCS reveals patterns of assault in which setting is a crucial differentiator not only of the people involved but of outcomes. The solid basis of McClintock's (1963) typology is reinforced, though some amendments to his categories are indicated. In particular a class of violence related to work should be introduced which has rather different parameters from domestic and particularly pub settings. Moreover, a need is indicated for a rather fuller reflection on situational criteria than McClintock allows. The complementary effects of the perceptual qualities of place should be set alongside its behavioural and physical properties. A typology of environmental settings of violence may then be enunciated that will be relevant in the search for more effective strategies for avoiding or preventing violence and indeed perhaps coping with it as it arises.

Much work remains to be done. Brantingham and Brantingham (1981; 1984) provide a model for the offender's use of environmental cues in their criminal activities. Can victims employ similar strategies in reverse to help cope with the prospect of violence or with confrontations as they arise? It is clear from the patterns evident in the BCS data that no single nor simple avoidance strategy can be developed encompassing all likely situations but rather a bundle of parallel and location-specific criteria that draw on particular relationships in particular contexts and which may be targeted at the victims most likely to be at risk in a given situation. For some settings, the environmental approach has already begun to tease out a framework of lessons to be learnt (Ramsey 1982; Walmsley 1986; Zimring and Zuehl 1986). If in other areas darkness remains, the task is clearer.

Acknowledgements

The author is indebted to the Home Office Research and Planning Unit for making available data from the British Crime Survey and specifically to Mr M. Hough for technical advice. The cross-tabulations and calculations presented here are, however, the author's own. The ACORN residential classification was developed by CACI Ltd for the statistical clustering of enumeration districts from the 1981 Census.

References

Block, R. (1977) Violent Crime: Environment, Interaction and Death, Lexington, M.A.: Lexington Books

Bottomley, K. and Coleman, C. (1981) Understanding Crime Rates, Aldershot: Gower

Brantingham, P.L. and Brantingham, P.J. (1981) 'Notes on the geometry of crime', in Brantingham, P.J. and Brantingham, P.L. (eds) Environmental Criminology, Beverly Hills: Sage

Brantingham, P.J. and Brantingham, P.L. (1984) Patterns in Crime, New York: Macmillan

Clarke, R.V.G. (1980) 'Situational crime prevention: theory and practice', British Journal of Criminology 20, 136-45

Davidson, R.N. (1981) Crime and Environment, London: Croom Helm

Davidson, R.N. (1982) 'Micro-environments of violence: situational factors in violent crime', unpublished paper, IBG Crime and Space Conference, London, May, 1982

Davidson, R.N. (1986) 'Micro-environments of assault: the role of location in violent injury', in Herbert, D.T., Evans, D.J., Davidson, R.N., Smith, S.J. and Mawby, R.I., The Geography of Crime, Occasional Paper No. 7, Department of Geography and Recreation Studies, North Staffordshire Polytechnic, 24-32

Gottfredson, M.R. (1981) 'On the etiology of criminal victimization', Journal of Criminal Law and Criminology 72, 714-26

Gottfredson, M.R. (1984) Victims of Crime: the Dimension of Risk, Home Office Research Study No. 81, London: HMSO

Heal, K. and Laycock, G. (eds) (1986) Situational Crime Prevention: from Theory into Practice, London: HMSO

Hindelang, M.J., Gottfredson, M.R., and Garofalo, J. (1978) Victims of Personal Crime: An Empirical Foundation for a Theory of Personal Victimization, Cambridge, Mass.: Ballinger

Hough, M. and Mayhew, P. (1985) Taking Account of Crime: Key Findings from the 1984 British Crime Survey, Home Office Research Study No. 85, London: HMSO

Hough, M. and Sheehy, K. (1986) 'Incidents of violence: findings from the British Crime Survey', Research Bulletin 20, 22-6

Jones, T., MacLean, B., and Young, J. (1986) The Islington Crime Survey, Aldershot: Gower

Lambert, J.R. (1970) Crime, Police and Race Relations, London: Oxford University Press
McClintock, F.H. (1963) Crimes of Violence, London: Macmillan
Monahan, J. and Klassen, D. (1982) 'Situational approaches to understanding and predicting individual violent behaviour', in Wolfgang, M.E. and Weiner, N.A. Criminal Violence, Beverly Hills: Sage
Normandeau, A. (1969) 'Robbery in Philadelphia and London', British Journal of Criminology 9, 71-9
NOP (1985) 1984 British Crime Survey: Technical Report, London: NOP Market Research Ltd.
Poyner, B. (1981) 'Crime prevention and the environment', Police Research Bulletin 37, 10-18
Poyner, B. (1983) Design Against Crime, London: Butterworths
Ramsey, M.N. (1982) City-centre Crime: the Scope for Situational Prevention, Research and Planning Unit Paper No. 10, London: Home Office
Stokols, D. (1979) 'A congruence analysis of human stress', in Sarason, I.G. and Spielberger, C.D. (eds) Stress and Anxiety (vol. 6), Washington, DC: Hemisphere Press
Walmsley, R. (1986) Personal Violence, Home Office Research Study, No. 86, London: HMSO
Zimring, F.E. and Zuehl, J. (1986) 'Victim injury and death in urban robbery: a Chicago study', Journal of Legal Studies 15, 1-40

Chapter Five

GEOGRAPHICAL ANALYSES OF RESIDENTIAL BURGLARY

David J. Evans

Objectives

It may be stated that the objective of this type of analysis is twofold. Firstly it aims to assess the risk of particular households within British cities being subjected to residential burglary and secondly it considers ways in which this particular source of anxiety may be alleviated. Ultimately it is assumed that a particular household's risk of residential burglary may be assessed with some degree of precision in terms of a number of physical and social factors. This is a long-term objective and current geographical research into residential burglary is still at the stage of defining these factors and exploring ways in which they can be measured.

This paper is organized in three main parts. First, a case is made for a focus on research into residential burglary within the context of the geography of crime; this case is related to a number of key points of rationale. Second, the process of data production is examined and, third, five ways of studying residential burglary are identified and evaluated. Concluding comments are then made.

Why focus on residential burglary?

A first point of rationale is the variation in burglary rates between different neighbourhoods. As Davidson (1984, 61) states: 'Burglary is a phenomenon with a very uneven geographical distribution. Great disparities exist between communities in the extent to which they suffer from this problem'. Geographical analyses are, therefore, pertinent though it should be acknowledged that this kind of spatial variation is also true of many other types of crime.

Davidson (1984) specifies five hypotheses which might explain geographical variations in residential burglary. The affluence hypothesis suggests that wealthier neighbourhoods might attract more burglaries; the vulnerability hypothesis suggests that 'where the built environment lacks certain properties, notably clear demarcation of public and private space and ease of informal surveillance, the opportunities for some types of crime are increased' (Davidson 1984, 62); the social cohesion hypothesis suggests that areas with weak social cohesion, perhaps associated with lower levels of occupancy, suffer more crime; the reputation hypothesis suggests that a community with a 'rough' reputation will attract more crime; and the proximity hypothesis highlights the tendency of offenders to 'seek victims on the fringes of their own neighbourhood or just into the next one' (Davidson 1984, 64). These hypotheses have varying status (see, for example, Herbert and Hyde 1985) but all deserve further research testing by geographers.

A second point of rationale is that the calculation of risk-related residential burglary rates is relatively straightforward compared to some other crimes. Thus household information such as number and types of dwelling units, household composition etc. derived from the census may be used as a denominator for the calculation of burglary rates for enumeration districts. For other crimes, such as criminal damage or even assault, comparable data for use in the calculation of risk-related rates are much more difficult to establish.

The study of residential burglary is further prompted by the growth in officially recorded cases. This amounts to an increase of 141.9 per cent in England and Wales between 1973 and 1983. Some writers such as Hough (1984) have argued that this increase in officially recorded cases is more related to the growth in the reporting and the recording of cases of residential burglary than to an actual increase in the number of cases. He also shows that the growth in recorded burglaries involving loss in England and Wales between 1972 and 1981 was in the order of 112,000 incidents or a growth of 68.3 per cent, whilst the increase revealed by victim surveys of recorded and unrecorded incidents was 30,000 or a growth of 8.1 per cent. Reasons for the increase in reportage of such incidents may be related to increased insurance levels by householders against burglary loss and the greater ease of contacting the police through the use of telephones. The increase in police recording may be related to improvements in the standard of record-keeping (Hough 1984, 20, 21).

A fourth point of rationale is the level of fear about residential burglary in the population and its impact upon victims. The first British Crime Survey (Hough and Mayhew 1983, 24) suggested that burglary caused most anxiety amongst those respondents who expressed any kind of worry about crime (1) whilst the second survey (Hough and Mayhew 1985, 72) placed burglary third behind rape and sexual harassment as the offence causing the most worry about crime. As far as the impact of burglary is concerned, Maguire (1980) in a survey of 322 victims of residential burglary in the Thames Valley police force area between 1977 and 1979 found that 65 per cent of the respondents felt that it was still having some effect upon their lives four to ten weeks after the event. Mainly the impact was 'a general feeling of unease of insecurity and a tendency to keep thinking about burglary' (Maguire 1980, 264). The impact upon 25 per cent of the victims was more 'acute' and 'considerable' at the same four- to ten-week interval after the event. Similarly, the second British Crime Survey (Hough and Mayhew 1985, 70) showed that burglary created more emotional problems than the other crimes included in the survey. (2)

A final point of rationale for the study of residential burglary is the relatively low clear-up rates of cases which amounted to 25 per cent in England and Wales in 1984 (Home Office 1985, 40). Each of these points of rationale underlines the significance of residential burglary as a form of criminal offence; it is a type of offence which can be linked to the residential environment.

Production of burglary data

In 1984 burglary in a dwelling represented 13.6 per cent of all notifiable offences recorded by the police in England and Wales (Home Office 1985, 28). In the production of such statistics at least two key decisions are involved. First, the decision to report to the police a case of residential burglary and second, the decision by the police to record the incident. Previous research has shown that it is primarily the victim or neighbours who report cases of residential burglary and that the police have a very minor role in their reportage. For example, in a study of Newcastle-under-Lyme Evans and Oulds (1984, 345) found that in a sample of data spanning four years, the victim reported on average 74 per cent of the cases, neighbours reported on average 13.7 per cent of the cases and the police 1.1 per cent of the cases. Decisions by members of the public are, therefore, of critical importance in the reportage of cases of residential burglary.

Victim surveys enable researchers first to assess the level of reportage of cases of residential burglary and second to discover if there are any systematic variations in who is likely to report such an incident. In comparison with other crimes the level of reportage of cases of residential burglary is relatively high. For example, the first British Crime Survey (Hough and Mayhew 1983, 11) indicated a reportage of 66 per cent of all cases of residential burglary, the second British Crime Survey an overall reportage level of 68 per cent, with a level of 87 per cent for burglary with loss and a level of 51 per cent for attempted and no loss burglaries (Hough and Mayhew 1985, 61, 21). The General Household Survey (OPCS 1982, 81) indicated a reportage level of 78 per cent of all cases of burglary and theft from private households. In a crime survey conducted in seven contrasting areas of Sheffield (Bottoms et al. 1987, 145) the reportage rate for burglary was 68 per cent. Such a relatively high reportage level for cases of burglary is not, however, found in the Sparks et al. (1977, 157) victim survey based upon interviews in three areas of London: Brixton, Hackney, and Kensington. In this case, for the crimes of burglary and theft in a dwelling, the average reportage rate was 51 per cent.

The main reasons cited for the non-reportage of household offences in the first British Crime Survey were that the incident was too trivial (49 per cent of the cases of non-reportage) and that the 'police could do nothing' (34 per cent of the cases of non-reportage) (Hough and Mayhew 1983, 11).

It should be noted that victimization surveys also have specific drawbacks. A first problem which may be identified is that of coverage in that usually only some crimes rather than all crimes are included in such surveys. The kinds of crime normally included are those with a clearly identifiable victim such as street crime. The problem of coverage also includes those of sampling since not all households in any given area can be interviewed. The second problem is that of comparability since both the respondent and the interviewer may interpret the question and the response in different ways. The 'halo' effect may also be evident in the sense that the interviewer may be seen as 'demanding' details of victimizations. Some surveys such as the second British Crime Survey use a two-stage interview procedure, initially checking off crime incidents and later recording their details. This can be seen as taking the pressure off the respondents. The third problem of victimization surveys is that of memory effects. Are crimes memorable incidents? Skogan (1986, 87) quoting Biderman noted that 'Respondents have to do a great deal of thinking and slow reflection before they can remember even fairly serious crimes of which they were victims some time ago'. A second aspect of memory effects is the possibility of lying by respondents about, for example, non-stranger assaults and a third aspect is the telescoping of victimization incidents which may involve bringing forward 'old' victimizations into the survey's reference period. The results of a study of Schneider which compared survey and police data are quoted by Skogan (1986, 93). The study found that 'on the average, matched incidents were pulled forward within the period by 2.2 months. Forty-nine per cent of all incidents were placed in the wrong month by their victims'.

A final problem of victimization surveys is the different productivity of respondents. As Skogan (1986, 95) notes 'in general, more highly educated respondents are more co-operative, more at ease in interview situations, and more able to recall the details of events'.

With regard to the production of official records the factors influencing the decision to report a burglary event may be divided into two categories. First the circumstances of the crime itself, especially its severity, and second the personal characteristics of the victim such as their age, sex, social class, ethnic group, and attitude to the police. Since several of these measures, especially ethnicity and social class are bases for residential segregation it is important that there is no systematic variation in reporting of burglary incidents by age, sex, social class or ethnicity. If there were any systematic variation in reporting by these measures any geographical variations in residential burglary would reflect these reporting tendencies rather than any other factors.

Generally the literature suggests that for all crimes 'it is primarily what happens to the victim rather than who is victimized that determines whether the victimization is reported to the police' (Hindelang 1976, quoted in Mawby 1979, 17). This kind of general trend is also found in the victim survey by Sparks et al. in three areas of London where they state that 'the most important factors influencing that decision (to report crime to the police) are primarily "incident-specific" and depend much more on the features of the particular situation than on the characteristics of the victims or their general attitudes or beliefs' (Sparks et al. 1977, 120).

With regard to the personal characteristics of the victims, for age, both the General Household Survey (OPCS 1982, 81) and the second British Crime Survey (Hough and Mayhew 1985, 67) show no significant variation in reportage of cases of burglary by the age of the victim. The social class of the victim does, however, have a more significant relationship with the reportage levels of cases of residential burglary, since the General Household Survey (OPCS 1982, 81) shows that heads of households in the higher socio-economic groups are more likely to report cases than those in lower socio-economic groups. This same trend is apparent in data drawn from the second British Crime Survey, where the differences in reportage levels of the most serious crimes against property (which included burglary) by social class were statistically significant (Hough and Mayhew 1985, 24). Both Hindelang (reported in Mawby 1979, 17) and Sparks (1977, 118) suggest that female victims tend to report more crimes than male victims. The evidence with regard to ethnic variations in reporting behaviour is ambiguous. In Smith's (1982) study in Handsworth and Aston in Birmingham it was found that for all incidents the Asian sample had the highest reportage rate of 53.8 per cent of incidents compared to 35.3 per cent for the 'other' category, which included the white respondents, and 21.1 per cent for West Indian respondents. However Tuck and Southgate (1981, 21) found in a victimization study in inner Manchester that the West Indian sample reported 80 per cent of all break-ins they experienced compared to a white reportage level of 68 per cent. The influence upon reporting behaviour of the offender rate of the area in which the victim lives and of tenure is illustrated by part of the Sheffield study (Bottoms et al. 1987, 145). Offender rate has little impact upon victims' reporting behaviour since the mean reporting rate for high offender rate areas is 30 per cent (for all crimes) and the corresponding figure for low offender rate areas is 31 per cent. The victim reporting rates for the different tenure areas are 27 per cent for all crimes for the local authority rented areas, 37 per cent for all crimes for the private rented areas and 22 per cent for all crimes for the owner-occupied area. A final factor which may influence the reportage of cases of residential burglary is attitudes to the police. The second British Crime Survey (Hough and Mayhew 1985, 67) found no significant relationship with this characteristic and the reportage of the most serious crimes against property, which included burglary cases. However, Smith (1982, 236) in explaining the under-reportage of crime incidents by West Indians in parts of inner Birmingham said that 'the tendency for West Indians to under-report is less surprising, given the deterioration of their relationships with the police in recent years'.

The second important decision in the creation of official burglary statistics is the decision by the police to record reported cases. Bottomley and Coleman (1981, 68) state that the reasons for not recording crime complaints are, first, in the case of mistakes or false reports, second, if complaints fail to meet legal requirements, third, the exercise of police discretion and fourth, if the complainant does not wish to proceed with the complaint. With regard to residential burglary Maguire (1982, 12) states that 'although the gatekeepers of police stations - the desk sergeants - may deflect many complaints which contain potential crimes on the grounds of ambiguity, triviality, or mischief-making by the complainant, reports of break-ins are treated seriously and are normally referred immediately to the CID.'

Police recording levels of cases of residential burglary are relatively high. Thus the second British Crime Survey (Hough and Mayhew 1985, 61) indicated a recording level of 70 per cent for burglary cases. McCabe and Sutcliffe (1978, 55, 56) in a study of police reaction to crime complaints in Salford and Oxford indicate a recording level of 76 per cent and 60 per cent respectively for alleged cases of breaking and entering, (3) whilst Sparks <u>et al.</u> (1977, 157) indicate an average recording level for reports of burglary and theft in a dwelling of 40 per cent for the three areas of London which they surveyed. Levi and Jones (1985) in a study concerned with public and police perceptions of crime seriousness, compare the rankings of a series of crimes which were graded in terms of their severity by a socially-mixed sample of the public and a sample of police officers. For the offence, 'the offender breaks into a person's house and steals property worth £20', whilst the public sample ranked this offence tenth in the series, the police sample ranked it fifth. This suggests that the police view this particular crime more seriously than members of the public and this may lead to higher police recording levels of this crime.

Whilst the reporting and recording of cases of residential burglary is not fully comprehensive, it appears that in relation to other crimes residential burglary is in a relatively favourable position with regard to these two particular aspects of data production. This, together with the particular problems of victimization surveys, suggests that officially recorded data for residential burglaries may usefully be utilized in geographical analyses.

Spatial and ecological studies

Research into the patterns of residential burglary within British cities suggests that two types of area have high rates. These are some parts of the inner city and some public-sector housing areas. These kinds of area are identified as being most at risk in four studies (Maguire 1982, 37), Herbert (1982, 58), Evans and Oulds (1984, 346), and Bottoms et al. (1987, 139). For example, Maguire (1982, 37) states, with reference to the distribution of burglaries in Reading in 1978, that 'the greatest concentration occurred in residential areas close to the town centre and in the large council housing estate to the south. The middle-class areas of the north escaped comparatively lightly'.

However, Maguire (1982, 37) also notes that 'in the centre of town, Reading burglars selected a disproportionate number of high-value properties'. Herbert (1982, 58) specifies, with respect to the distribution of residential burglary and thefts from dwellings in West Swansea in 1975 that 'the correspondence (of these crimes) is not with prime "target" areas of the wealthier suburbs' but with 'the edge of the central commercial area' and 'some of the large public-sector estates'. In the Sheffield study (Bottoms et al. 1987, 139) the highest burglary rates were in the survey areas which were council housing with a high offender rate and private rented housing with a high offender rate.

Evans and Oulds (1984) construct three maps depicting

different residential burglary rates within Newcastle-under-Lyme for the period 1978 to 1981. The overall burglary rate shows eight enumeration districts with high rates, six of these are described as 'almost exclusively council-owned estates' and two as 'decaying areas of older owner-occupied housing'. These kinds of area are also emphasized in the distribution of lower-value and higher-value burglaries. It is only in the latter category that some middle-class housing areas appear as high-rate areas. In addition Evans and Oulds (1984, 350) calculate the proportions of the social area clusters, of low socio-economic status, having relatively high residential burglary rates. The clusters with the highest proportion of enumeration districts with high rates are the two types of area described above, which both have 32 per cent of their enumeration districts with high rates. This illustrates the fact that only parts of these social area clusters may be characterized as having high residential burglary rates. Within the limits of ecological analyses the principal associations found for residential burglary in the Newcastle-under-Lyme study were with low socio-economic status, local authority rented housing, and enumeration districts with a child-rearing or adolescent age structure (Evans and Oulds 1984, 351).

Journey to burgle

Analyses which investigate characteristics of the journey to burgle, whether measured in terms of physical or social distance, offer a further avenue of enquiry. Such analyses rely on data for burglary cases where an offender is apprehended. As for offence data and perhaps to a greater degree, such data do have limitations. Firstly such data are not comprehensive since nationally not more than one third of all cases of residential burglary are 'cleared'. Secondly it has been shown by Burrows and Tarling (1982, 20) that a third of burglary cases may be cleared as offences 'taken into consideration'. Some researchers such as Lambert have suggested that the clear-up rate was, therefore, 'very dependent upon the whims of offenders declaring their interests in prevailing exploits' (Lambert 1970, 43). Thirdly, Davidson (1984, 68) has shown that clear-up rates may vary spatially being higher in 'council' than 'inner-city' or 'owner-occupied' areas; the respective rates for a sample of 78 cases in Hull being 32 per cent, 12 per cent and 14.5 per cent. The Newcastle-under-Lyme study showed the proximity of offender and victim in physical terms since the distance frequencies of 131 burglary trips show that 48.9 per cent of the burglaries were committed within 0.8 km of the offender's home and that only 13.9 per cent were committed more than 5 km from the offender's home (Evans and Oulds 1984, 352). This kind of relative proximity is also demonstrated for 'breaking' offences by Baldwin and Bottoms' (1976, 83) study in Sheffield, where they found that 74.8 per cent of such offences which were cleared, were committed within two miles of the offender's home.

If one examines proximity in terms of social rather than physical distance, a social area typology may be used as a framework for assessing social proximity of victim and offender. Such a framework was employed in the classification of 73 cleared cases of residential burglary committed between 1 January 1980 and the early part of 1982 in the Stoke-on-Trent South police force area. (4) The residential location of victim and offender were related to a six-fold social area typology produced from the 1981 Small Area Statistics for the City of Stoke-on-Trent (Jackson and Oulds 1984). The analysis showed that 56.8 per cent of the 101 burglar-victim interactions took place within similar types of social area, principally within cluster type 6 (council) and within clusters 1 and 2 (inner city). In this sample of data the main 'exporter' of burglars to other types of social area was cluster type 6. A similar type of analysis conducted by Davidson (1984, 65) relating to 141 victim-burglar interactions within the City of Hull found that 'some 60 per cent of residential burglars confine their activities to the same sort of areas in which they reside' and that in this case it was the inner-city areas rather than the council housing areas which were the main exporting areas of burglars.

Whilst relative physical and social proximity between offender and victim may be shown from these types of analysis, as far as the reasons concerning target selection are concerned, at best only inferences may be made. A more direct approach with regard to the decision-making of offenders is that which adopts behavioural techniques.

Behavioural approach

A behavioural approach would examine the decision-making underlying burglary offences, especially with regard to the selection of targets (see also Chapter 8 by Rengert in this text). It assumes that such a decision-making process is rational and involves at least two geographical decisions. First, the selection of a particular neighbourhood within which to select a target and second, the selection of a particular house. A distinction has to be made between young opportunistic burglars who take presented rather than sought opportunities for burglary and older professional burglars who seek opportunities for burglary.

Systematic behavioural analyses which have been conducted, such as those, for example by Maguire (1982) and Bennett and Wright (1984), have been concerned with the decision-making processes of the latter category of offender. (5) Their analyses point to the importance of 'planning' in the commission of burglary offences. Thus Maguire (1982, 81) in interviews with forty men in custodial institutions said that 'the vast majority of burglaries described to us were planned'. Bennett and Wright (1984, 43) in their behavioural surveys classified 6.8 per cent of the offences as 'opportunistic' with no time gap between the decision to commit a burglary, the selection of the target and the commission of the offence, 47 per cent as 'search' offences where a time gap occurred between the decision to burgle and the selection of the target and offence commission and 59 per cent as 'planned' offences where a time gap occurred between the selection of a target and offence commission.

With regard to 'cues' guiding the selection of the target for burglary, Bennett and Wright divide the factors involved in the choice into three: the perceived risk of getting caught; the reward; and the ease of entry. Of these it is the first which is most important in influencing the selection of the target and the most important of the risk factors are surveillability and occupancy (Bennett and Wright 1984, 93). Surveillability includes such factors as: the degree of cover; whether the property was overlooked; the availability of access from the front to the rear; the proximity of neighbours; distance from the road and the presence of passers-by. With regard to occupancy 'over 90 per cent of the offenders who took part in the semi-structured interview said that they would be either conditionally or unconditionally deterred by the presence of occupants' (Bennett and Wright 1984, 94). Occupancy proxies, of which the two most important are burglar alarms and dogs, also had a powerful deterrent effect. Maguire's (1982) analysis is less systematic in its analysis of the factors influencing the selection of targets for burglary. However, among the factors mentioned in his survey are: familiarity with the area; occupancy; escape routes; cover; and the presence of burglar alarms.

In addition to asking convicted burglars which factors influenced their choice of target, inferences may also be made by comparing selected characteristics of victim and non-victim households. This is the type of approach which was adopted by Winchester and Jackson (1982) in their study of burglary in Maidstone, Tonbridge, Malling and Sevenoaks in Kent in 1979. Again as with Maguire (1982) and Bennett and Wright (1984) they say that the burglaries in this suburban area are likely to be 'the more lucrative, professional burglaries' (Winchester and Jackson 1982, 5). They compare the relative importance of four factors in influencing the incidence of burglary victimization: environmental risk; occupancy; reward; and security. They conclude that 'the most important factor ... proved to be environmental risk, followed by occupancy rates and reward in this order. Relative security levels did not contribute to this discrimination' (Winchester and Jackson 1982, 22). The four main factors comprising environmental risk were: distance from the nearest house; distance from the road in which the house stands; majority of sides of house not visible from public areas; and access at both sides of the house from front to back on the plot (Winchester and Jackson 1982, 39).

A further study which makes inferences about the factors influencing the choice of targets by burglars by comparing a sample of victimized households with a sample of non-victimized households is that by Evans (1987). This study is concerned solely with burglary in an affluent housing area and compares a sample of 111 households which were victimized between 1.1.1984 and 31.10.1986 and a randomly drawn sample of households who were not victimized during the period for which data were collected. The aspects of the households chosen for comparative purposes were the natural area location, the rateable values, specific physical characteristics, and certain details of their micro-location. Findings were that the burgled sample was more accessible in a variety of ways and was more affluent. Of the three natural areas with the highest overall burglary rates, two were located along major routeways and the other contained particularly large dwellings. The rateable value comparison showed that whilst the mean rateable value of the burgled sample was £269.10p, the mean value of the control sample was £226.71p and the frequency distribution of the two samples across the range of rateable values was significantly different in that the burgled sample had a greater representation in the higher rateable value categories. (6) Two ways in which the physical characteristics of the two samples varied were that more of the burgled sample were on major routeways and had clearly visible access from front to rear of the plot. (7) Finally more of the detached houses in the burgled sample had at least one open side or were located on a corner. (8)

These behavioural analyses have pointed to the importance of surveillance, occupancy, accessibility, and relative affluence, in influencing burglary victimization.

The defensible space approach

Newman (1972, 3) defines defensible space in the following way. 'Defensible space is a surrogate term for a range of mechanisms ... that combine to bring an environment under the control of its residents. A defensible space is a living residential environment which can be employed by inhabitants for the enhancement of their lives, while providing security for their families, neighbours and friends'. Four major categories of defensible space characteristics were identified: territorial influence; surveillance opportunities; the capacity of design to influence the perception of a project's uniqueness, isolation and stigma; and the influence of geographical juxtaposition with 'safe zones' on the security of adjacent areas (Newman 1972, 50).

Several writers such as Mawby (1977) have criticized this concept on several grounds. In general terms, two main areas of debate may be identified. First, the emphasis on physical or architectural features as opposed to social characteristics in explaining offence rates. This emphasis has been modified in Newman's more recent writings (Newman 1976) and the distinction can never be a mutually exclusive one, except in a most artificial way. Second, the ambiguity of the elements of the concept. Some of these ambiguities have been described by Mawby (1977, 176). More specific criticisms have been aimed at Newman's (1972, 39) methodology in establishing the validity of the defensible space concept. In particular, his uncritical acceptance of officially recorded offence information, the lack of offender data for his matched pair of housing areas and the fact that in the social data he provides for Van Dyke and Brownsville, some differences are apparent in the levels of residential mobility in the two areas and in the possession of assets by residents of the two areas (Newman 1972, 250). Nevertheless as Poyner (1983, 36) suggests 'it is reasonable to assume that the ideas of increasing residents' territorial control ... and encouraging natural surveillance would be relevant to burglary in so far as they discourage access to building ... by intending burglars'.

Herbert (1982, 63) compares some of the surveillance characteristics of a sample of 228 burgled dwellings with those of an equivalent number of non-burgled dwellings. The significant differences between the two samples were that the burgled dwellings had fewer back or side doors and fewer back gardens. The former would reduce surveillance opportunities and the latter would improve access to the dwelling. In the Newcastle-under-Lyme study (Evans and Oulds 1984, 349) two samples of data were plotted at the intra-neighbourhood level. In each case 62 per cent of the victimized households were adjacent to space, usually public space, which allowed rear access to the dwellings. Surveillance and accessibility are, therefore, characteristics which have found empirical support in influencing the incidence of residential burglary.

With regard to future research upon residential burglary Newman's four original defensible space hypotheses may be reformulated in the following terms. The influence of territoriality could be broadened to include measures of accessibility to victimized households. This characteristic was important in distinguishing victimized households in Evans' (1987) study of burglary in an affluent housing area (see p. 98) and in the study of burglary in Newcastle-under-Lyme described above. The importance of surveillance has been alluded to by Bennett and Wright (1984, 93), by Herbert (1982, 63) and is incorporated in Winchester and Jackson's (1982, 39) concept of environmental risk. This concept is one, however, that needs a fuller empirical base since the specific importance of this characteristic has not been fully tested in a direct manner in a variety of social localities.

The concept of image is again a concept capable of fuller development. The fact that burglars in an affluent housing area do select more affluent properties suggests that at least in that context image is important. In addition if this characteristic is broadened to include social images then characteristics which enhance the vulnerability of problem estates may be addressed. Finally with regard to milieu, the vulnerability of corner locations and of dwellings adjacent to open space has already been referred to (p.98). This is also a concept capable of development in a more varied way with respect to the incidence of residential burglary.

Social hypothesis

The social hypothesis considered here is that of the relationship between residential burglary and social cohesion.

The influence of social cohesion is inconsistent even within individual studies. Thus Reppetto (1974, 47, 62) finds a strong negative relationship between social cohesion and the rate of residential crime at the aggregate scale but not such a clear relationship at the scale of the individual household. Similarly Herbert and Hyde (1985, 271) in their study of residential crime in Swansea state that:

> neighbourhoods which scored highly on the indices of social cohesion and interaction employed in the study included both high and low offence-rate areas. The vulnerability of the two local authority estates to residential crime could not be explained in terms of low social cohesion within the estates. It was only really in the inner city that the contrasted offence rates between the two sample areas seemed to owe anything to differing levels of local social control.

This does not mean that a lack of social cohesion has no relationship with the incidence of residential burglary but that this particular association is relevant in only some types of burglary. It is a complex rather than a simple relationship which needs further investigation.

On a more micro-scale Walsh (1980, 107) has demonstrated the importance of neighbour relations in influencing the incidence of residential burglary. Thus with regard to his victim sample of sixty-seven households he says 'the disturbing feature was that in 79 per cent of cases the householder had poor or non-existent relationships with his immediate neighbours on both sides of his house, and in only eleven cases was the householder friendly with immediate neighbours on both sides'. This kind of detail is rarely referred to in studies of residential burglary and does point to a significant factor especially given the popularity of neighbourhood watch schemes.

Conclusion

In summarizing the material presented so far the risk factors may be divided into three categories; the first category includes the standard structural attributes of the household such as social class and tenure, the second the physical or architectural features of the household and the third social or cultural attributes of the household.

In terms of location the areas most at risk from residential burglary are either inner-city areas or parts of the local authority sector (see Gottfredson 1984, 21 and the spatial studies noted earlier: p.93). At a more localized scale corner positions, locations adjacent to open space and proximity to a major routeway are risk factors (see Evans and Oulds 1984, 349 and Evans 1987). A final locational factor which increases burglary risk is proximity to offender areas. This characteristic is important because of the relative proximity of the residential location of offender and victim noted earlier (p. 95). For the social class of victims whilst the first British Crime Survey suggests no relationship between residential burglary and this attribute (Hough and Mayhew 1983, 19), the General Household Survey of 1980 shows a higher risk for households of lower socio-economic status (OPCS 1982, 80). There is also an indication in the literature that within any particular area households of higher relative affluence will be selected as burglary targets (see Maguire 1982, 37; Maclean et al. 1986, 3.13.; and Evans 1987). With regard to the type of accomodation and tenure the households most at risk, are those which reside in flats or which are of private rented or local authority tenure (see the 1980 General Household Survey OPCS 1982, 80; and the first British Crime Survey, Hough and Mayhew 1983, 18). Gottfredson's (1984, 24) analysis of the data collected by the first British Crime Survey and also Walsh's (1980, 101) suggest that small rather than large households are more at risk from residential burglary.

With regard to physical or architectural features the attributes which increase the risk of residential burglary are low levels of occupancy, poor surveillance characteristics, and specific features such as clear access from front to rear of the building plot. Occupancy and surveillance characteristics were the two key characteristics which Bennett and Wright (1984, 93) identified as increasing the risk of burglary. Surveillance characteristics are also a large component of Winchester and Jackson's concept of environmental risk (Winchester and Jackson 1982, 39). The specific physical risk factor of clear access from front to rear of the house has been identified by Walsh (1980, 106), Bennett and Wright (1984, 93) and Evans (1987).

The social and cultural factors include social cohesion both at the neighbourhood level and relations with neighbours. Studies reviewed earlier (p. 100) show that some ambiguity surrounds the former's relationship with residential burglary whilst one study has demonstrated the importance of the latter. With regard to the levels of cultural organization on an area basis Bottoms and Xanthos (1978) contrast the sub-cultural organization of the high offender rate housing area which they study with the social disorganization of the high offender rate flat area which they study. These findings have some parallels with those of Herbert and Hyde (1985) showing that only in some areas is social and cultural cohesion associated with the incidence of residential burglary.

An alternative way of classifying risk factors is to categorize burglaries into two or three types and if possible to relate these particular types of residential burglary to specific types of geographical area. For each a set of physical, economic, and social variables could be devised and ranked in order of importance as contributing to the incidence of a particular type of burglary. Three types of burglary may be identified, each associated with a particular geographical area (after Davidson 1984, 68). Inner-city burglaries take place in areas with low social cohesion, low occupancy levels by some groups, poor security, and poor surveillance opportunities. In this case the burglars are locally based. In the public sector housing area burglaries are associated with strong social cohesion and with economic deprivation. In this case the burglar and the offender may know each other and the youth of a particular area may be the burglars. In affluent areas, factors such as environmental risk, occupancy, and low social cohesion appear to be associated with the incidence of residential burglary. Proximity to 'offender' area and newness of residential areas may also be risk factors. The burglars involved are non-local.

Whichever way of categorizing risk factors is adopted analysis remains to be done in identifying and, if possible, quantifying the risk factors involved in residential burglary. In particular the importance of characteristics such as accessibility, surveillance, image, milieu, and social cohesion upon the incidence of residential burglary needs to be worked out in a variety of empirical situations so that our understanding of the reasons for geographical variations in burglary may be further enhanced.

Acknowledgements

The author acknowledges the research assistance provided by North Staffordshire Polytechnic in conducting burglary studies in Newcastle-under-Lyme and Stoke-on-Trent. The assistance of the Chief Constable of Staffordshire, Mr C.H. Kelly and his officers in conducting such studies is also gratefully acknowledged.

Notes

1. The percentage of 'worried' respondents mentioning each type of crime were:

burglary	44%
mugging	34%
sexual attacks	23%
assaults	16%
vandalism	4%
vehicle theft	1%

(Hough and Mayhew 1983, 24)

2. The other crimes included in this part of the survey were:
theft of vehicle
theft from vehicle
robbery, attempts, snatch theft
wounding

(Hough and Mayhew 1985, 70).

3. Comparable recording levels for the other crimes included in the survey were:

	Salford	Oxford
disturbance	11%	7%
domestic	11%	2%
damage	30%	50%
assault	50%	29%

(McCabe and Sutcliffe 1978, 55, 56)

4. This analysis was based upon original charges which were cleared up rather than cases cleared by other means.

5. Bennett and Wright (1984, 9) say 'compared with all offenders convicted of burglary, those in our sample were slightly older than average' and 'that as a result of selecting predominantly from offenders in custody the groups were more experienced than the typical convicted burglar'.

6. The rateable values of the two samples were as follows:

Value in pounds	1		2	
less than 150	6	(5.4%)	5	(4.5%)
151-200	31	(27.9%)	14	(12.6%)
201-250	40	(36.0%)	41	(36.9%)
251-300	20	(18.0%)	24	(21.6%)
301-350	8	(7.2%)	13	(11.7%)
351 or more	6	(5.4%)	14	(12.6%)

Chi square test = 11.279 with 5 degrees of freedom significant at the 5 per cent level.

column 1 = non-burgled sample
column 2 = burgled sample

7. The data for the two samples with respect to position on major routeways and access from front to rear of the house are as follows:

On what type of road does the house stand?

	1		2	
private/pedestrianized	1	(0.9%)	3	(2.7%)
limited access/ cul-de-sac	45	(40.5%)	27	(24.3%)

two way road off main road	52	(46.8%)	63	(56.8%)
'A' road/through road	13	(11.7%)	18	(16.2%)

Chi square test = 6.34 with 2 degrees of freedom (category 1 was omitted) - significant at the 5 per cent level.

Is there clearly visible access from front to rear of the house on:

Left side		1		2
yes	44	(39.6%)	61	(55.0%)
no	67	(60.4%)	50	(45.0%)

Chi square test = 4.626 with 1 degree of freedom significant at the 5 per cent level

Right side		1		2
yes	33	(29.7%)	37	(33.3%)
no	78	(70.3%)	74	(66.7%)

Chi square test = 0.187 with 1 degree of freedom not significant at the 5% level

column 1 = non-burgled sample
column 2 = burgled sample

8. Whilst 32 (53.3 per cent) of the detached houses in the burglary sample have at least one open side, 16 (27.5 per cent) of the detached houses in the control sample have a similar attribute.

10 (16.6 per cent) of the detached houses in the burglary sample occupy corner positions compared to 4 (6.9 per cent) in the control sample.

There were 60 detached houses in the burgled sample and 58 in the control sample.

References

Baldwin, J. and Bottoms, A.E. (1976) The Urban Criminal, London: Tavistock

Bennett, T. and Wright, R. (1984) Burglars on Burglary, Aldershot: Gower

Bottomley, A.J. and Coleman, C.A. (1981) Understanding

Crime Rates, Aldershot: Gower

Bottoms, A.E. and Xanthos, P. (1981) 'Housing policy and crime in the British public sector', in Brantingham, P.J. and Brantingham, P.L. (eds) Environmental Criminology, Chapter 11, 203-26, Beverly Hills: Sage

Bottoms, A.E., Mawby, R.I., and Walker, M.A. (1987) 'Localized crime survey in contrasting areas of a city', British Journal of Criminology, 27, 2, 125-54

Burrows, J. and Tarling, R. (1982) Clearing Up Crime, Home Office Research Study Number 73, London: HMSO

Davidson, R.N. (1984) 'Burglary in the community - patterns of localization in offender-victim relations', in Clarke, R.V.G. and Hope, T. (eds) Coping with Burglary - Research Perspectives on Policy, Chapter 5, 61-75, Boston: Kluwer-Nijhoff

Evans, D.J. and Oulds, G. (1984) 'Geographical aspects of the incidence of residential burglary in Newcastle-under-Lyme, UK,' TESG, 75, 5, 344-55

Evans, D.J. (1987) 'Burglary within an affluent housing area', unpublished research note

Gottfredson, M.R. (1984) Victims of Crime: the Dimensions of Risk, Home Office Research Study number 81, London: HMSO

Herbert, D.T. (1982) The Geography of Urban Crime, Harlow: Longman

Herbert, D.T. and Hyde, S.W. (1985) 'Environmental criminology: testing some area hypotheses', Transactions I.B.G., new series, 10, 3, 259-74

Home Office (1985) Criminal Statistics - England and Wales, 1984, London: HMSO

Hough, M. (1984) 'Residential burglary: a profile from the British Crime Survey', in Clarke, R.V.G. and Hope, T. (eds) Coping with Burglary - Research Perspectives on Policy, Chapter 2, 15-28, Boston: Kluwer-Nijhoff

Hough, M. and Mayhew, P. (1983) The British Crime Survey - First Report, Home Office Research Study No. 76, London: HMSO

Hough, M. and Mayhew, P. (1985) Taking Account of Crime - Key Findings from The Second British Crime Survey, Home Office Research Study, No. 85, London: HMSO

Jackson, G.A.M. and Oulds, G. (1984) A Social Area Analysis of Stoke-on-Trent, Occasional Paper No. 4, Department of Geography and Recreation Studies, North Staffordshire Polytechnic

Lambert, J.R. (1970) Crime, Police and Race Relations,

London: Oxford University Press
Levi, M. and Jones, S. (1985) 'Public and police perceptions of crime seriousness in England and Wales', British Journal of Criminology, 25, 3, 234-50
McCabe, S. and Sutcliffe, F. (1978) Defining Crime - A Study of Police Decisions, Oxford University, Centre for Criminological Research, Occasional Paper number 9
Maclean, B.D., Jones, T. and Young, J. (1986) Preliminary Report of the Islington Crime Survey, Centre for Criminology, Middlesex Polytechnic
Maguire, M. (1980) 'Impact of burglary upon victims', British Journal of Criminology, 20, 3, 261-75
Maguire, M. (1982) Burglary in a Dwelling, London: Heinemann
Mawby, R.I. (1977) 'Defensible Space - A Theoretical and Empirical Appraisal', Urban Studies, volume 14, p. 169-79
Mawby, R.I. (1979) Policing the City, Farnborough: Saxon House
Newman, O. (1972) Defensible Space, New York: Macmillan
Newman, O. (1976) Design Guidelines for Creating Defensible Space, Washington: US Department of Justice
O.P.C.S. (1982) General Household Survey: 1980, London: HMSO
Poyner, B. (1983) Design against Crime - Beyond Defensible Space, London: Butterworths
Reppetto, T. (1974) Residential Crime, Cambridge, Mass.: Ballinger
Skogan, W.G. (1986) 'Methodological issues in the study of victimization', in Fattah, E. (ed.) From Crime Policy to Victim Policy, Basingstoke: Macmillan
Smith, S.J. (1982) 'Race and reactions to crime', New Community, 10, 2, 233-42
Sparks, R.F., Genn, H.G., Dodd, D.J. (1977) Surveying Victims, Chichester: Wiley
Tuck, M. and Southgate, P. (1981) Ethnic Minorities, Crime, and Policing, Home Office Research Study number 70, London: HMSO
Walsh, D. (1980) Break-ins: Burglary from Private Houses, London: Constable
Winchester, S. and Jackson, H. (1982) Residential Burglary: The Limits of Prevention, Home Office Research Study, number 74, London: HMSO

Chapter Six

DISPOSITION AND SITUATION: TWO SIDES OF THE SAME CRIME

Alice Coleman

The mainstream in contemporary crime prevention is a trend away from the dispositional theory of criminal behaviour to a situational theory. Cornish and Clarke (1987) have attributed the change to the failure of psychiatric treatment to modify dispositional drives, combined with the discovery of crime-control potential in target hardening and other situational measures. They consider the main mechanism of crime to be a rational choice made by the offender in response to opportunities perceived in any given situation, with other contributory factors falling into place in a complex model.

Unfortunately, complex arguments evolved in high-echelon corridors of research often become blurred as they filter down to the implementation work-force, and the situational crime-prevention theory is no exception. A recent speaker at an Institute of Housing meeting summarized the concept as follows: 'If a flimsy door is kicked in, replace it with a stronger one. That's common sense and all there is to it.'

This oversimplification performs a useful function in clarifying the polarity of the dispositional versus situational stance. It assumes that the door is solely at fault for provoking the passer-by to exercise the rational choice of breaking into the premises, and evades the question of why a hundred earlier passers-by did not make the same rational choice. Why was the door a non-provoking situation to all of these and then suddenly transformed into a provoking situation? Trasler (1987) has suggested that some people have made a standing decision to commit crimes where they can, so that their rational choice consists of assessing the situation with crime in view. Others have a standing decision not to commit crime, and the difference can perhaps be considered dispositional.

The present paper argues that dispositional and situational aspects of crime are not mutually independent, but two sides of the same coin. This becomes comprehensible if we cease to regard a criminal disposition as an innate drive and think of it as a learned attitude picked up at an early age in the same type of crime-prone environment that subsequently maximizes opportunities for crime by rational choice.

The fact that the average age of the British criminal is only 14.7 years appears to restrict the major origin of delinquency to two possibilities. Either it is a genetic drive acquired before birth or else it is a learned response acquired during early childhood. If a genetic cause were the sole or predominant explanation, we should expect the incidence of criminality to remain roughly stable from one generation to the next, which does not fit the facts of the escalating crime rate as these are consistently revealed by official statistical returns. Environmental situations or cultural attitudes can change much more rapidly and therefore a learned disposition towards crime seems more likely.

There are many possible reasons why children reared since World War Two may have learned to be less law-abiding than their parents. The prevalence of violence on television, the effect upon the brain of lead in petrol fumes, the unhealthy nature of junk food, the temptation of open displays in shops, the softness of the law towards criminals under 16, the decline of religion. All these and other changes in society have some credible claim to be considered and are not discounted here, but the central theme discussed here is what, on the strength of research evidence, is put forward as a major factor in both dispositional and situational motivation for crime. This factor is Britain's massive demolition of traditional homes for non-traditional redevelopment as flats, where different households have to share the same buildings and grounds. Shared residential space appears to provide an early-learning environment capable of undermining normal child-rearing practices and promoting child crime on a scale formerly found only in the worst tenement slums - the 'rookeries' of British experience. Architectural situations that are highly vulnerable to crime can teach children to adopt criminal decisions, and this learned disposition can then cause them to see all situational weaknesses as rational opportunities for crime. Thus, disposition and situation both appear to be components of a unified design-disadvantagement theory, which argues that design improvement (DI) is a preventive measure that affects both simultaneously.

Methodological approach

Design-disadvantagement research was launched in 1979 under the sponsorship of the Joseph Rowntree Memorial Trust. Its findings were published as a book, Utopia on Trial (Coleman et al. 1985) which was aimed at a general readership and needs supplementary methodological explanation here.

The research aimed to test:

(i) Whether Oscar Newman's Defensible Space theory (1972) applied to Britain
(ii) Whether any design variables other than those which he nominated had a similar deleterious effect
(iii) Whether they promoted more types of anti-social behaviour than those he demonstrated
(iv) Whether there could be a more comprehensive system of design improvement to maximize the alleviation of social breakdown.

Originally six types of social breakdown were used as test measures; however, suitable crime data were not available until the book was in the press and most of the crime results have not yet been published.

Analytical methods are restricted by the nature of the data. Many of the design variables do not have interval values and none is normally distributed, which makes it inappropriate to correlate them with crime or social breakdown measures by means of Pearson's Product Moment method, or to establish the trend of a relationship by means of regression analysis. Non-parametric methods have therefore been substituted: Kendall's tau C correlation and trend lines. For data where Pearson's and Kendall's methods are both valid, the latter yields substantially lower coefficients, and people accustomed to Pearson's values may be inclined to dismiss them without using the standard safeguard of significance levels. Another factor which

Table 6.1 Residential crime in 729 blocks of flats

Design variable	Correlation total crime
Dwellings per block	0.437
Dwellings per entrance	0.349
Storeys per block	0.351
Flats or maisonettes	0.180
Overhead walkways	0.095
Interconnecting exits	0.173
Lifts and staircases	0.263
Corridor type	0.296
Entrance position	0.274
Entrance type	0.127
Stilts or garages	0.226
Gates or gaps	0.286
Play areas	0.163
Spatial organization	0.226

lowers the coefficient without impairing its value is the use of a very large number of cases spread over a large area, as this gives scope for more disturbing factors to operate. Table 6.1 gives results for 729 blocks of flats in the Carter Street Police Division of Southwark, which may be compared with 285 used by S. Wilson (1980) and only 59 cases by T. Hope in Burglary in Schools (1982). All the coefficients in Table 6.1 are very highly significant.

Disturbing factors that reduce the coefficients include the designs themselves. There are virtually no identically designed blocks and many different combinations of good and bad values. Some strongly effective positive designs offset other strongly effective negative designs in the same block, and hence low coefficients may result from the power of design rather than from its weakness.

For reasons such as these, correlation is regarded as a minor method and more reliance is placed on trend lines. These show the average change in the dependent variable in relation to differences in the independent variable, despite the effect of all other factors, whether known or unknown. While correlation reveals how far the other factors exert a disturbing effect, trend lines show the magnitude of change,

Figure 6.1 Scattergrams of crime levels plotted against design values

(a) A perfect correlation combined with a flat regression line, affording little scope for crime reduction through design improvement
(b) A low correlation coefficient combined with a steep regression line affording greater scope for crime reduction, as seen in (c)

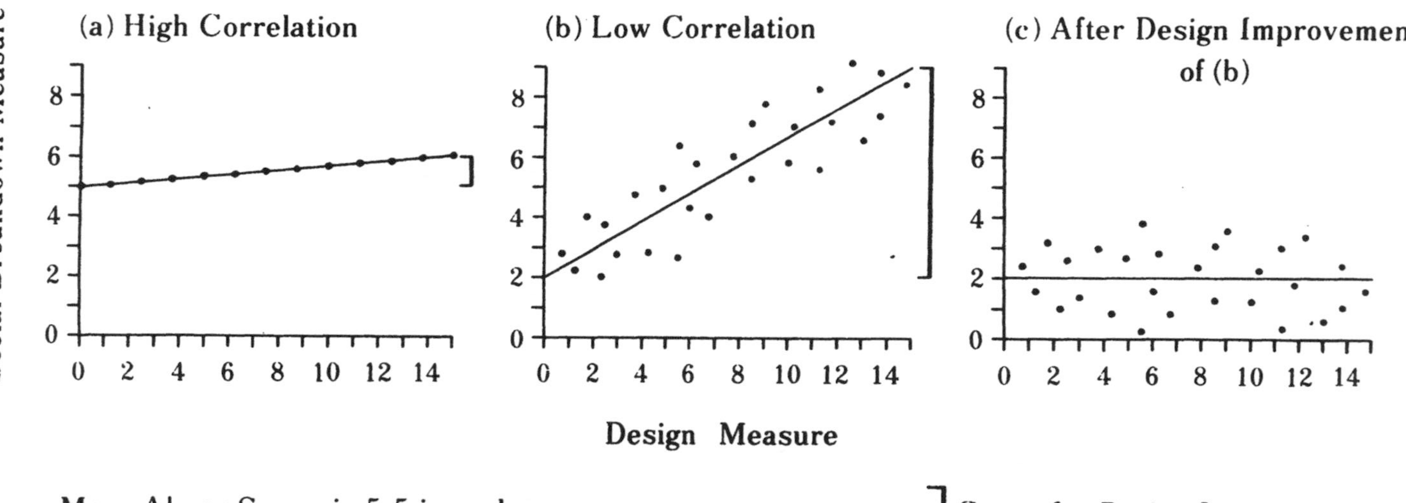

Figure 6.2 Trend lines for the percentage number of blocks having nine categories of crime for values 0 to 5 walkways per block

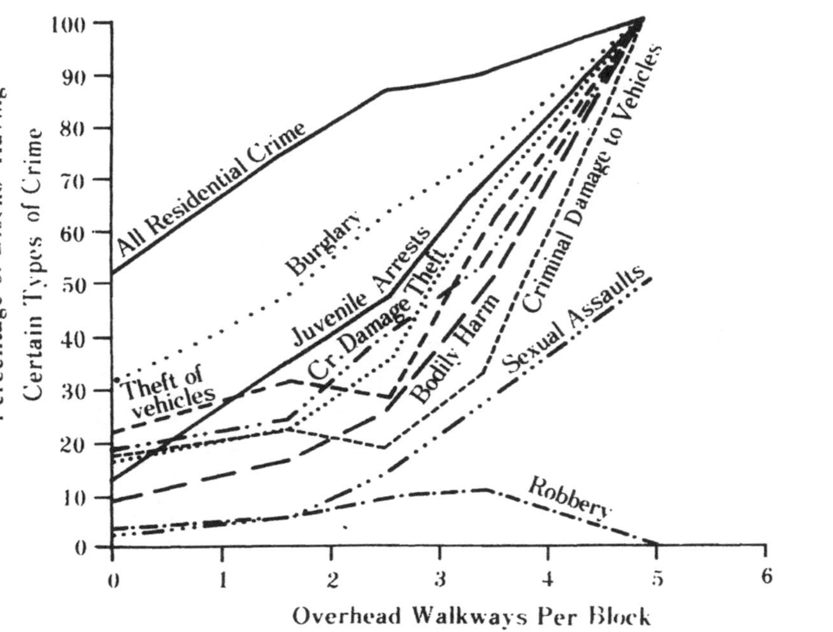

The curve labelled All Residential Crime also includes a few cases of drug offences. The apparent discrepancy in the robbery curve is an artefact due to the fact that there were only 29 offences in this category

which is of much greater practical use. The difference is not always understood. Hope, for example, has asserted that correlation and regression are the same thing (1986), and so Figure 6.1 is introduced to clarify the point.

Figure 6.1(a) shows a perfect correlation of +1.0, without disturbing factors. This might be hailed as a remarkable finding, but it would have no practical use. The regression line is so flat that any investment in changing the independent variable would produce only a tiny effect upon the dependent variable. Figure 6.1(b), by contrast, shows a low correlation coefficient, with widely scattered points, and a steeper regression line, and in these circumstances, investment would be well worth while. Figure 6.1(c) indicates the improvement to be expected if all the values of the independent variable were modified to equal its best value. The regression line would flatten and the scatter of points would change. This could represent a marked drop in crime, and it shows why trend lines are a more important source of information than correlation coefficients. Sometimes the two give widely different results. For example, trend line evidence shows that overhead walkways are the most powerful factor in the spread of crime to the maximum number of blocks, but correlation evidence suggests that it is only thirteenth in order of importance.

Overhead walkways

Figure 6.2 shows slightly smoothed trend lines for crime categories plotted against the number of overhead walkways joining each building. The dependent variable is the percentage number of blocks having at least one crime. This percentage increases very rapidly as the design value goes from 0 to 5 walkways.

Because walkways appear to have such a powerful effect, their removal is potentially a useful crime-prevention device, a prediction that has been confirmed by the demolition of a few of them on two Westminster Estates. In Lisson Green the crime rate was halved, and although Poyner (1986) found it subsequently began to rise again, it did not nearly approach the level predictable by extrapolating the pre-demolition figure in accordance with London's generally rising trend. In the Mozart Estate, burglary rates were compared for the five-month periods before and after demolition, to reveal a drop of 55 per cent, and even more encouraging was a small decline in the surrounding area, which rules out any appreciable displacement.

It is not enough to block up walkways instead of removing them. Utopia on Trial (Coleman et al. 1985) referred to cases where vandalism increased two or three times after blockage. This follows from the fact that children penetrate the barrier while adults are held back and child behaviour worsens as there is no restraining adult presence. A recent blocking-off exercise in a Greenwich estate brought a temporary reduction in litter and graffiti, but two barriers were soon attacked and a relapse seems likely.

Number of storeys

Figure 6.3 shows that the trend lines for building height are more irregular than those for overhead walkways but these irregularities are useful as they indicate the presence of disturbing factors worth investigating. This is an advantage which trend lines have over regression lines.

Regression lines are always grossly smoothed to straight lines or simple curves, whereas trend-line smoothing can be regulated. The aim here has been minimal smoothing to remove irregularities due to small numbers of cases while retaining more important fluctuations. The method was running means of three classes, or where that yielded fewer than 50 cases, four classes.

The most striking irregularity in Figure 6.3 is the tailing off of the curves for juvenile arrests and children-in-care in blocks of over 12 storeys. This does not imply that juveniles are more law-abiding in tall buildings but reflects the general recognition of high-rise designs as bad for children leading to social workers' efforts to have problem families moved out. However, although progressively higher blocks have fewer juvenile arrests, they also have progressively more crimes of each sort, suggesting that they encourage lawlessness in previously non-criminal families.

Another irregularity in Figure 6.3 is the fact that each line either flattens or dips near five storeys before resuming its upward trend. This, too, reflects a child-related policy. Blocks of five storeys or more normally have lifts, and because tiny children cannot operate them, local authorities have tried to avoid placing families with pre-teen children in vacant dwellings above this level. But children grow older

Figure 6.3 Trend lines relating the number of storeys per block to all residential crime and nine separate crime categories

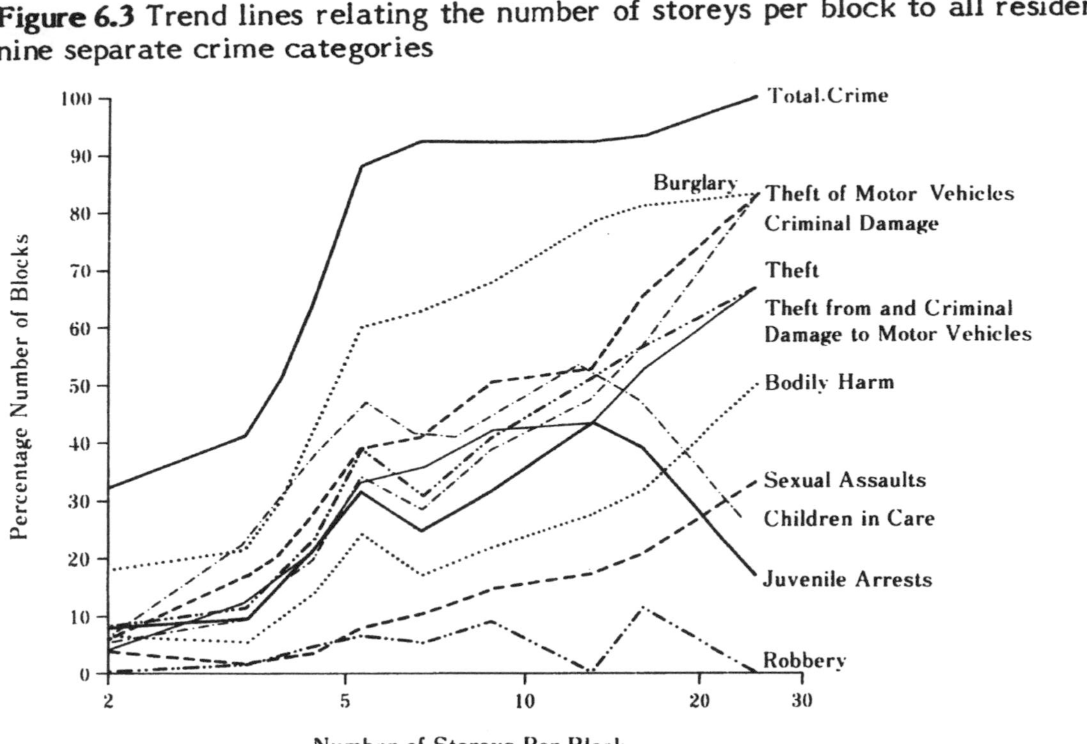

Note: For explanation of irregularities see text.

and soon learn to use the lifts, ascending in noisy joy-rides and leaving a trail of obscene graffiti all down the stairways. Fewer children may cause relatively less damage, but in absolute terms their havoc continues to increase with the height of the block.

Wilson and Sturman (1976) found that child density was a somewhat stronger factor in vandalism than four design variables and this gave impetus to the drive to disperse families with children (see also Wilson 1980). Design was neglected because it was thought density reduction would be a solution in itself, and show that design had a negligible effect. In practice, however, there has been a reverse outcome. By stripping away the disturbing factor of child density, the role of design has been revealed more clearly, and a study of 44 wards in Utopia on Trial showed that its correlation coefficient had become twice as strong as that for child density.

Sunderland housing authority developed a method for improving blocks of flats by 'top-downing' them into single-family houses with gardens. This has proved extremely popular with local residents, and the policy has been replicated in other areas where declining populations make the sacrifice of dwellings possible. The effects of this policy upon crime rates deserves close examination. Top-downing is less appropriate in the south-east with its severe housing shortage, but there are some very perversely designed estates where it may be well justified.

Disadvantagement scores

Overhead walkways and high rise are design disasters that have reached public consciousness, but thirteen others were identified in Utopia on Trial. They will not be discussed individually because practical improvement requires an understanding of their combined effect.

The combined measure is the disadvantagement score, which is the total number of defective designs coexisting in the same block. 'Defective' means that the value of the variable breaches the threshold level at which various forms of abuse exceed statistical expectation (Utopia on Trial, p. 125 and Appendix I). When the disadvantagement score is plotted against the percentage of blocks having crime (Figure 6.4), a remarkable picture emerges. In Figures 6.2 and 6.3 even the best values are associated with appreciable crime percentages, as other designs with more harmful

Figure 6.4 Trend lines relating the disadvantagement scores of 729 blocks of flats to the presence or absence of nine types of crime and all residential crime (ARC)

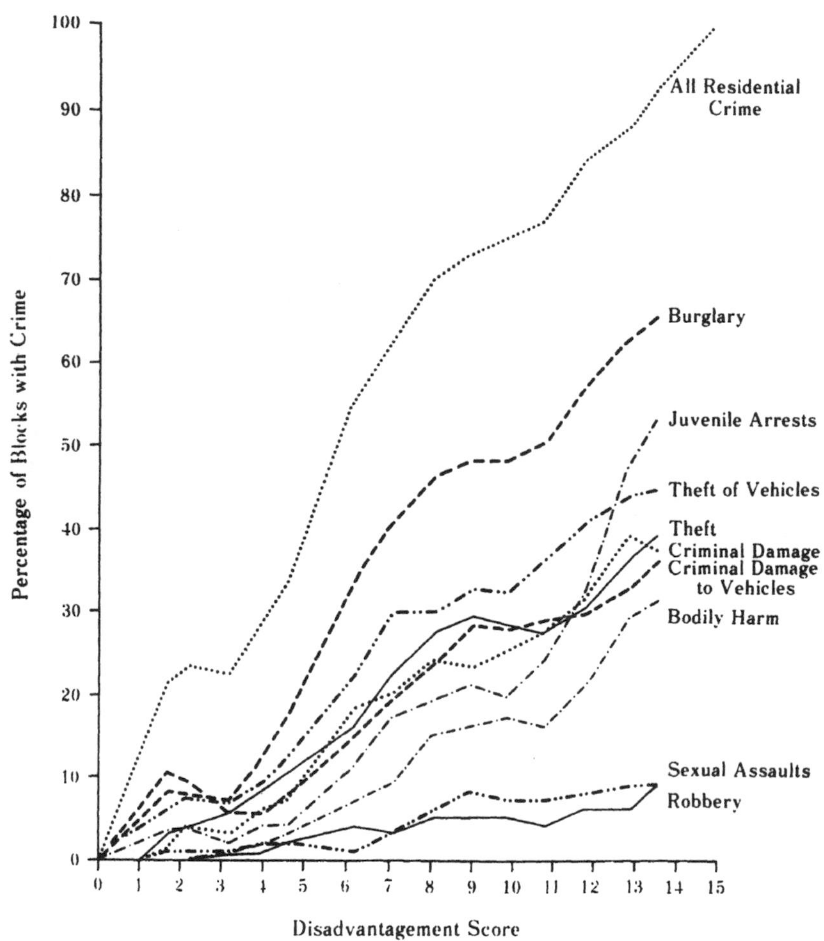

Note: The curve for juvenile arrests, which is independent of resident reporting, rises more steeply than those for other crimes after a score of 10, i.e. in blocks where it is suspected that unreported crime is greatest.

values are also present in some of the blocks, but the disadvantagement score of zero isolates those blocks where no variable breaches its threshold and this group did not report a single crime during the study year. The curve for total crime rises diagonally across the graph to culminate at 100 per cent for the maximum score of 15.

The practical message of this graph is the need to try reducing disadvantagement scores by design improvement schemes, in the hope that there would be a correspondingly larger percentage of blocks left crime-free. The greater the number of design variables that can be brought down to their thresholds, the greater the promise of crime reduction.

A per-dwelling basis

Before dealing with the practical aspects of design modification, it is necessary to respond to an academic criticism of the use of blocks rather than dwellings. It is argued that blocks with worse scores are inevitably larger and that a proportional increase in crime is only to be expected, regardless of design. It is further claimed that design is therefore irrelevant.

This hypothesis has to be tested by plotting crime on a per-dwelling instead of a per-block basis, and also by plotting actual numbers of crimes instead of mere presence or absence in each block. Figure 6.5 attempts this by using two trend lines for comparison. The continuous line represents the average number of dwellings per block and confirms that there is, as expected, a general tendency to increase as the disadvantagement score worsens. The broken line represents the average number of crimes per block, which also increases. The test lies in ascertaining whether the crime curve rises in proportion to the size curve, as the critics allege, or whether it rises at a faster rate, in which case design is associated with a crime increase over and above a purely pro rata size effect. The answer is mixed.

For scores 0, 1, 2, and 3, block size is roughly constant but the number of crimes increases. Between scores of 3 and 7 the increase is closely pro-rata, but from 7 to 15 size is more or less progressively outstripped by crime. Blocks with scores of 13, 14, or 15 average one residential crime per five dwellings per year, in addition to the non-residential dangers of a criminal neighbourhood.

Figure 6.5 The average number of dwellings and of crimes per block is plotted for each disadvantagement score, showing that the volume of crime increases more rapidly than block size as design worsens.

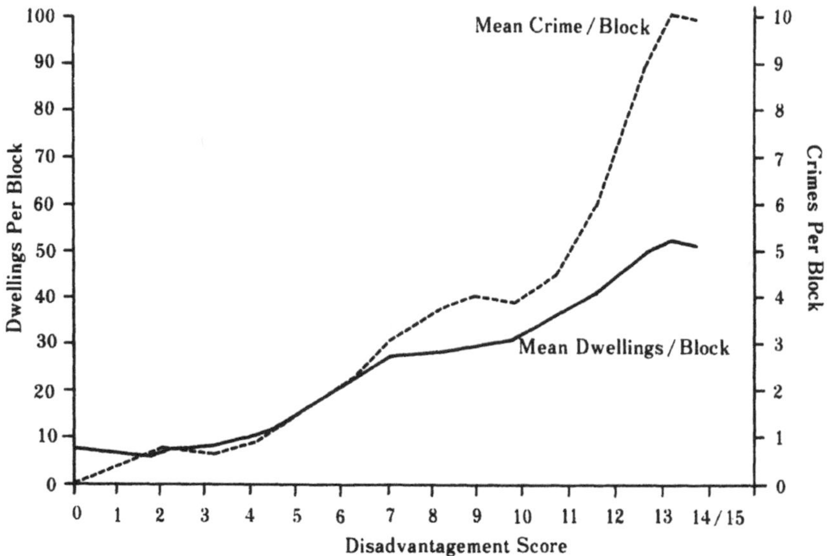

Some critics dismiss the evidence on the grounds that it takes no account of the dark area of unreported crime, but this is a double-edged argument, as there are several reasons for thinking that unreported crimes are commonest in the same areas as reported. The criminals themselves may be reluctant to approach the police if they become victims, as also may law-abiding parents with delinquent children, while honest citizens may be intimidated into silence by threats of reprisals. Supporting evidence is now afforded by the pattern of juvenile arrests shown in Figure 6.4. This is the one graphed category that is not committed in the block, but involves offences such as thefts from shops, for which the juveniles are arrested elsewhere, and plotted according to the design of their homes. The trend is independent of reporting from the blocks and the way in which it diverges from the other curves is therefore informative. In general it harmonizes with the fanshaped pattern from scores 0 to 10, but thereafter steepens to overtake four other categories and converge upon a fifth, and it is here that under-reporting of residential crime seems most concentrated.

The corollary is that the difference in per-dwelling crime rates between the best and worst designs is even greater than Figure 6.5 suggests and that a closer approximation would be obtained if juvenile arrests alone were plotted against average number of dwellings (Figure 6.6). This does, in fact, show a more marked acceleration, with no trace of a pro-rata increase over any part of the range. Juveniles in blocks with scores of 13, 14, and 15 seem to be seven to eight times as much at risk of having a criminal record as those in blocks scoring 0 or 1.

Even this may still underestimate the difference, as the worst scores include the tall towers from which families with children are being progressively excluded. If we could calculate juvenile arrests per juvenile, instead of per dwelling, the corrupting effect of misconceived design would be recognized as even more powerful. Conversely, the hope of crime reduction through design improvement is also more powerful.

Is that hope realizable or is the damage permanent? This cannot be resolved by further rarefied dispute but only by practical evidence. Design improvement needs to be tried and tested.

Practical trials

When a problem estate has been surveyed and disadvantagement scores established for each block, recommendations can be formulated. This is not always easy, as some estates are so perversely designed that a simple reversal of one defect may create another. It takes considerable background knowledge and ingenuity to find ways of reducing the maximum number of designs down to their threshold levels, thus lowering the score to the greatest possible extent.

It is also essential to work with the tenants. It is their home territory and they should have as big a voice as possible in decisions concerning it. The role of a design-disadvantagement survey is to provide them with scientific information to help them make their choices more wisely and this needs to be an ongoing process to negotiate any pitfalls that may arise between adoption of the recommendations and their implementation.

Figure 6.6 The average number of juvenile arrests increases very much faster than block size as design worsens, as there is no systematic bias in the under-reported component

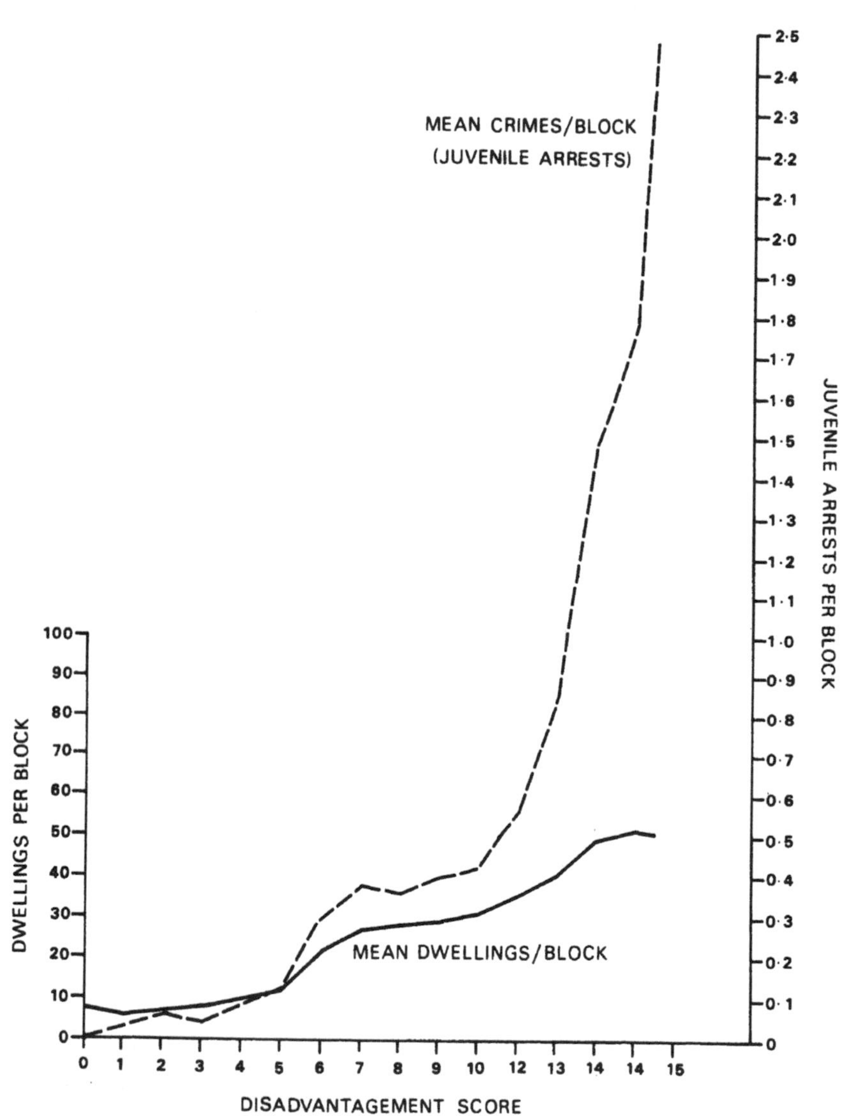

For over forty years the Department of the Environment and its predecessors have assiduously brainwashed planners, architects and the public with plausible arguments about the superiority of certain principles of design, and although these have now been proved counterproductive, it is not surprising to find that people's thinking tends to slide back down the old, well-worn groove as design-improvement details are crystallized. In the North Peckham Estate, for example, after the London Borough of Southwark had received a set of recommendations on how the disadvantagement score could be reduced from 13.1 to 3.9, they proceeded to introduce their own modifications and unwittingly introduced several major retrograde designs which raised the 'improved' score to about 8. This was described as 'better value for money', but the value would be merely reduction of crime to an average inner-city rate - vastly worse than the low crime conditions predicted by the original recommendations.

The case of the Mozart Estate is a complete contrast. Because Westminster Housing Department both accepted the recommendations and retained advice through the implementation stages, false trails could be nipped in the bud and an ingenious idea produced by an authority architect could be incorporated. The Tenants' Association was also a constructive influence, and the outcome is a planned reduction of the score from 12.8 to 3.9, instead of 4.8 as originally envisaged.

Although phase I of the Mozart scheme, with its 55 per cent drop in the burglary rate, is the only practical implementation to date, there are a number of estates where one or more design improvements have been incidental to other projects and all the known examples have proved encouraging. Walkway removal, which is the first step, has already been mentioned as having a direct effect upon crime and this would have the additional advantage of making riots easier to control.

The second step is to wall off each block separately, so that outsiders can no longer take short cuts through or past it. This was tried in the Brandon Estate in Southwark and although no before and after crime figures are available, the residents of the block concerned reported a greatly reduced fear of crime; two ground-floor families who had boarded up their windows for fear of breakage and burglary, took down the boarding and began to live in daylight again. The exclusion of strangers allowed residents to get to know each other more easily, decreasing anonymity. It also meant fewer litter-droppers and graffiti-scrawlers, so that tenants were able to control and eliminate these abuses and take a new pride in their premises. Some of them began to reclaim overgrown flowerbeds, thus introducing gardening as one answer to the estate's 'lack of activities'.

Intelligent enclosure of individual blocks can remedy no fewer than four deleterious designs. As well as creating one-site blocks it can ensure that there is only one perimeter gate, which like the communal entrance to the block, should face a public road. There should be no confused space left between the walls of adjacent blocks; if the distance is too great to divide it between them, new houses can be inserted to create a more continuous frontage and concentrate pedestrian movement along the road. This gives a bigger public presence, more security, and fewer opportunities for intruders to dodge round buildings.

The third step consists of adjustments to the buildings now safely enclosed in their own separate grounds. Some are still too large and anonymous and need to be divided into smaller self-contained sections. Westminster divided its Holcroft Estate into vertical sections, with fewer dwellings accessible from each entrance and on each corridor, and fewer lifts, staircases, and exits mutually accessible from each other. Although no variable was modified completely down to its threshold level, the amount of litter, graffiti and vandalism was curbed sufficiently to yield a substantial saving in maintenance costs. In Lewisham a case of sealing all alternative exits put a complete stop to vandalism, partly because the vandals were deprived of escape routes and partly because more concentrated use of the main entrance increased their fear of detection.

An alternative way of dividing a block is horizontal partitioning, as applied in Wandsworth's Surrey Lane Estate. All the ground-floor maisonettes were given individually fenced front gardens and a resident has described how this transformed the children from unruly menacing gangs to polite individuals. Previously they had called for their friends by banging on doors or windows as they passed, but the new type of entrance constrained them to open the gate, walk up the path, knock on the door and speak to the parents. The parents discovered who their children's friends were and addressed them by name when they saw them about the estate. This dispelled the youngsters' sense of anonymity and engaged them in normal conversation. It seems, therefore, that the lack of front gardens is a design villain contributing to the generation gap, and its influence begins very early. Where there are no individual fenced gardens, toddlers are deprived of an opportunity to learn respect for other people's property, and also for its named and known owners.

Constructive though Wandsworth's garden policy has proved, it is not the best value for money. If they could have had expert design-disadvantagement advice, they could have used the same funding to reduce the disadvantagement score by six points instead of one, with a correspondingly larger impact upon the crime rate.

The recent demand for a Minister for Play is a reminder of the propaganda that has presented play areas as a wholly beneficial sacred cow. Figure 6.7 tells a different story; blocks with play areas are more prone to every kind of crime, as well as having more children in care and juvenile arrests. St Helen's, Lancashire, had a problem estate that ceased to be a problem when the tenants spontaneously dismantled all the play equipment, and many estates furnish bitter tales of the conflict and hostility generated by play areas. Critics are often frustrated by the apparent lack of an alternative yet a solution does exist. Individual back gardens are safe places where toddlers can gradually develop self-confidence by going out alone at a crucial early stage, so that later they are more venturesome in exploring the wider neighbourhood and making use of parks and other areas where there is an adult presence. Flat-bred children do not develop the same independence, but retain an immature desire to hang around the home estate, even in their teens. The remedy would be to transfer families into ground-floor flats or into new houses built on the spare green space so copiously available. The Ferrier Estate in Greenwich, for example, has room for 400 extra houses, as well as a park.

Halton District, Cheshire, transferred families out of some of its upstairs maisonettes, which were then divided into small flats. This brought a noticeable reduction in litter and graffiti, even though the number of dwellings in the block had increased.

These examples cover all the recommended design changes except one; no test is yet known of the conversion into dwellings of garage space or areas between stilts on the ground level under raised blocks. All the others have had positive results. The best example to date is one that, by

Figure 6.7 Play areas related to the percentage of blocks with crime

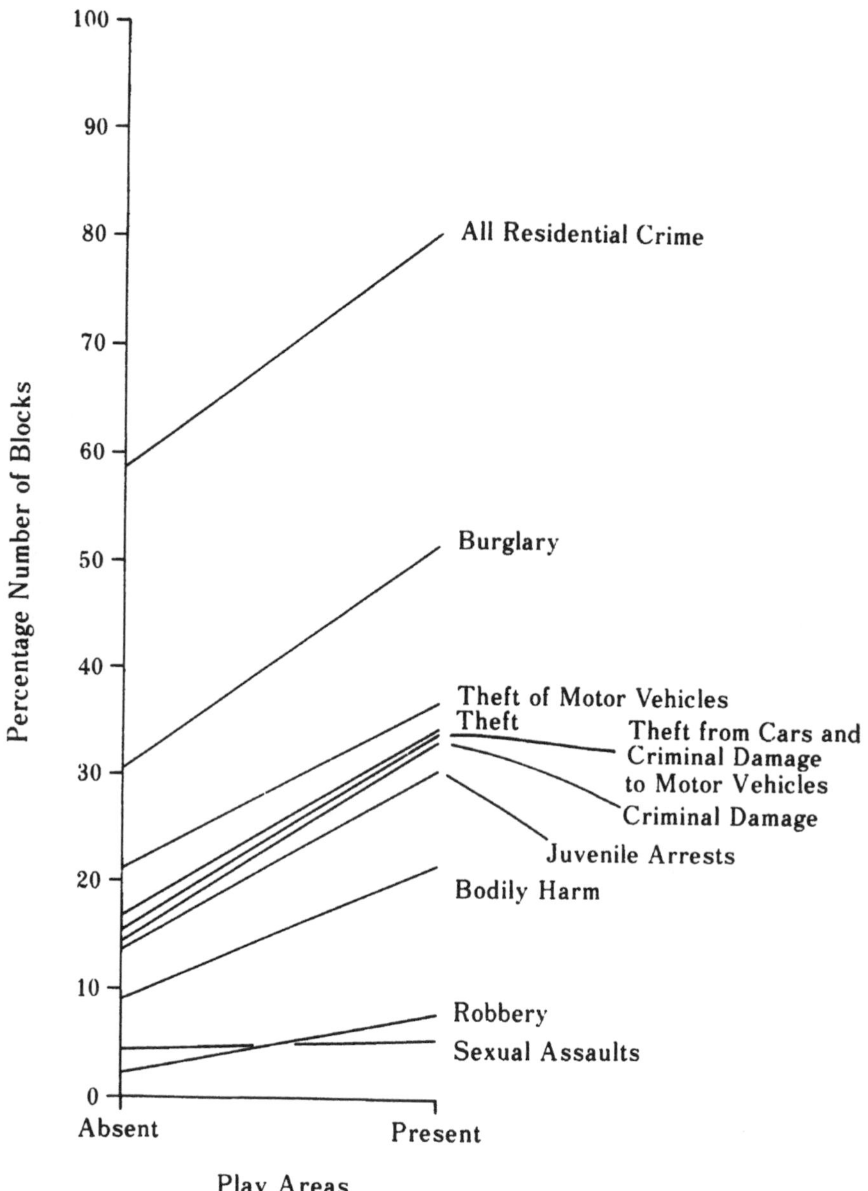

chance, has functioned as a controlled experiment, in Hackney. Lea View and Wigan House are two estates built to the same design in the same area at the same 1930s period, both have recently been refurbished to the tune of over £6 million each. Both were problem estates with high crime rates, but now the improved part of one has been crime-free for a few years while the other still has one crime per dwelling per annum. The crucial difference is that Wigan House made no design improvements while Lea View allocated 'a small fraction of the cost' (Thompson 1984) to improving seven relevant variables, of which five were brought right down to their threshold values.

The difference in value for money is profound. Wigan House soon relapsed to a filthy state of litter, excrement, graffiti, and vandalism, with all its hopeful relandscaping reduced to bare trodden earth. Lea View has retained its new pride in a litter-free, graffiti-free and vandal-free estate, with beautifully kept individual gardens. The tenants say they have thrown away their sleeping pills and no longer need the psychiatrist, while the fact that a formerly dangerous park opposite the estate is now regarded as safe and usable, shows that crime seems to have been cured rather than merely displaced.

Counter evidence

It is also important to know of any cases where design improvement has been tried and failed. So far there are no proven instances of this, although several false claims have been made. These spring from the loose use of the term 'design change' to mean any kind of alteration, and not the modification of just those fifteen variables that research has proved relevant. Design improvement (DI) is used in a technical sense to refer to scientifically supported changes only.

One of the mistaken concepts is the substitution, mentioned earlier, of walkway blockage for walkway removal. Another is the fact that locks, entryphones, etc. are often vandalized. These are not design improvements but security improvements (SI), and it is of interest that security research is following a path parallel to true design research. Both are recognizing the adverse effect of enforced sharing of premises or facilities and are working towards more autonomy for the individual household. Protection of the communal entrance is giving place to

Table 6.2 Relevant design features of houses

	Design variable	Beneficial values	Worst values
Facade			
1	Window shape	Bay (oriel, flush)	Recessed, none
2	Window type/position	Clear glass giving outlook when seated or standing	beaded or frosted, too high, too small, none
3	Door	Slightly recessed, flush	Projecting porch, garage, etc. sideways door
4	Intervisibility	Facades intervisible across a road	Radburn type (front facing back of next row)
Frontage			
5	Garden depth from facade to frontage	10-15 feet	Less than 10 feet, and greatly over 15 feet
6	Side fences between neighbours	Both sides (between adjoining doors, where appropriate)	One side, neither side, between door and garden of the same house
7	Front fence or wall	Waist-high; strong	Above eye level, knee-high, no fence, flimsy
8	Gate	Waist-high	None, high gate

(Table 6.2 continued)

	Spatial context		
9	Fronting	On to a public road	On to a path or estate road
10	End of road	Corner house with front garden facing both roads	End house with high faceless side walls
11	Back garden	Enclosed; no gate	Open or with gate
12	Abutting rear use	Back gardens of next road or enclosed land use	Path, road, open use

stronger doors and windows for each flat, where the responsibility of one family is not vitiated by the carelessness of another.

Poyner (1983) has criticized conclusions about DI in Hunter's (1978) report on the Angela Street Estate, Liverpool, where individual gardens for ground-floor flats proved successful while those for upstairs flats did not. In design-disadvantagement terms, this is exactly what would have been predicted. The plots reserved for upstairs tenants were 'confused space', and the fact that Wilson chose to label them 'semi-private' did not endow them with the true semi-private characteristics that are necessary for success. They did not directly adjoin their owners' individual dwellings with surveillance from a window and access from a door and they were fenced off from the dwelling just as strongly as from outsiders. They are far from being evidence that genuine DI does not work.

Houses

As well as recommending the modification of existing problem blocks, Utopia on Trial advocates that no further flats should be built. There are already more than enough for the people who really prefer them.

However, houses can also be vulnerable to crime, especially in the case of modern designs which ape the deleterious features of flats, e.g. shared lawns instead of individual gardens. Twelve disadvantaging designs have been identified, and unfortunately these are innovations which contemporary builders feel constrained to incorporate (Table 6.2).

A recent visit to a London Docklands Development Corporation estate revealed that every house had either seven or eight of these defects, and barely had a high burglary rate been predicted than a member of the local neighbourhood watch team revealed that two-thirds of the houses had been burgled, some more than once, during the estate's two-year life-span. The LDDC intends to impose even worse crime-prone designs in Newham, and its brash, uninformed attitude is allowed to override the protests of Newham Housing Department, which has been learning from earlier mistakes. The proposed multiplication of this type of development corporation in other inner cities could be a recipe for disaster, and it is hoped that the architectural liaison officers of the police will be able to give constructive guidance.

Scientific testing

The academic findings of design-disadvantagment research reveal a strong association between the design characteristics of houses and flats and the probability that they will be prone to various types of crime and social breakdown. Associations are not necessary causal, but in this case there are several independent indicators that a causational relationship does exist.

First, as with the link between smoking and cancer, the probability of causation becomes stronger as sample size is increased. Since Utopia on Trial was published the original data set of 4,099 blocks of flats has been expanded by approximately another 1,500, and many others have been observed, though not surveyed, in the company of local housing officers. In general, social breakdown indicators were recorded rather than crime figures, but the prevalence of crime was frequently referred to by officials, and an interesting case was provided by a cluster of slab blocks in the new town of Tomakomai in Hokkaido. These had disadvantagement scores of about 12, which led to a prediction that there would be crime there, even in the low-crime Japanese culture. Despite the assurances of the chief planner that he had heard of no crime, the Tenants' Association had in fact found it necessary to appoint a crime-prevention officer.

Secondly, there is historical evidence. It is fashionable at present to assert that crime is caused by unemployment but that does not explain why the great depression of the 1930s was a relatively law-abiding time, nor why the massive rise in crime began during the post-war period of affluence and full employment. The upwave-downwave pattern of unemployment is completely out of phase with the steady rise in lawlessness and must be ruled out as a cause. The growth in deleterious design, by contrast, is closely in phase, and cannot be ruled out.

Even when a cause is firmly established, there is no guarantee that its removal will reverse the effect, which may have taken on a life of its own. The fact that the association between crime and design does seem to be reversible in at least a few cases is an added reason for entertaining the idea of a causal link.

The time seems ripe, therefore, for a scientific programme of testing the hypothesis and it is suggested that this should be on the same scale as the Department of the Environment's Priority Estates Project to test the effect of management improvement (MI), in ten estates. The latter has not been very scientific. MI has been combined with SI, and also with RI (improvement through repairs, redecoration, refurbishment, relandscaping, etc., in an attempt to restore the estates to their pristine condition). There may also have been improved policing (PI), and the provision of community facilities (CI), and these various approaches have mingled in a way that obscures which of them may be responsible for any success.

DI should be applied on its own, to ascertain exactly what it can achieve and also exactly what it costs. Is it good value for money? Without the precise costing provided by proper tests, this question cannot be answered definitively, but there are many positive pointers which have led John Banham of the Audit Commission (1986) to say that DI would not be mere expenditure but investment with returns, and that Utopia on Trial should be required reading for all local authorities.

DI often seems to be a cheaper alternative in itself. Nationwide Building Society, for example, finds top-downing profitable, and Halton District estimated DI schemes for two estates as costing about one-third of the proposals they superseded.

DI also saves future costs. While MI involves the employment of extra staff in the hope that their salaries will be covered by letting vacant properties, design change is a once-for-all operation that does not necessitate extra staff costs in perpetuity and actively saves on maintenance because of the reduction in vandalism. The rents from increased letting would then be an added bonus.

DI is also expected to produce financial serendipities in fields other than housing. With fewer badly-behaved estate children in schools, there would be savings to be derived from less vandalism, theft, burglary and arson on educational premises, and educational standards would be easier to raise. There would also be a spillover into the cost of law and order, with fewer calls on police, probation officers, the courts and prisons. A third area affected would be the National Health Service, which is currently having to cater for a wide range of stress illnesses common in problem estates, but likely to disappear, as in the case of the Lea

View Estate, Hackney, when design disadvantagement scores are substantially reduced.

Design improvement, therefore, offers a large promise, and it seems only reasonable to finance proper scientific testing on a sufficient scale to ascertain its actual costs and savings in terms of crime and other forms of social breakdown.

Conclusion

The theme of this paper is that the dispositional and situational theories of criminal behaviour are not separate and opposite, but closely interlocking parts of a single, unified theory of design disadvantagement, in which the residential environment of early childhood exerts a strong situational influence, encouraging children to adopt standing decisions to commit crimes where opportunities offer. As design varies, in well documented ways, so, too, does the probable criminalized proportion of each estate-bred cohort of children.

The design-disadvantagement theory does not claim to explain all crime, but the statistical evidence suggests that it probably explains more than other factors. This offers hope of a very substantial reduction in crime if the theory can be turned to practical account. The theory also offers a further major ground for hope, as it places the blame on the built environment, which is more easily and permanently changed by design improvement than a mental bent towards crime can be changed by psychiatric treatment. It may even be that the latter can become more effective when the criminal patient is not released back into a criminogenic environment. The optimism inherent in the design-disadvantagement theory should not remain merely academic, but should be subjected to rigorous scientific testing to establish whether its promise is justified.

References

Audit Commission (1986), Managing the Crisis in Council Housing, London: HMSO, 34-7

Banham, J. (1986) Personal communication

Coleman, A. with Brown, S., Cottle, L., Marshall, P., Redknap, C. and Sex, R. (1985), Utopia on Trial,

London: Hilary Shipman

Cornish, D.B. and Clarke, R.B. (1987), 'Situational crime prevention, displacement of crime, and rational choice', in Heal, K. and Laycock, G. (eds) Situational Crime Prevention: From Theory into Practice, London: Home Office Research and Planning Unit, 1-16

Hope, T. (1982) Burglary in Schools: The Prospects for Prevention, London: Home Office Research and Planning Unit, Paper 11

Hope, T. (1986) Personal communication

Newman, D. (1972) Defensible Space New York: Macmillan

Poyner, B. (1983) Design Against Crime: Beyond Defensible Space, London: Butterworths, 53-6

Poyner, B. (1986) Personal communication

Thompson, J. (1984) Community Architecture: The Story of Lea View, Hackney, London: Royal Institute of British Architects Community Architecture Group

Trasler, G. (1987) 'Situational crime control and rational choice: a critique', in Heal, K. and Laycock, G. (eds) Situational Crime Prevention: From Theory into Practice, London: Home Office Research and Planning Unit, 17-24

Wilson, S. and Sturman, A. (1976) 'Vandalism research aimed at specific remedies', Municipal Engineering, May, 705-6

Wilson, S. (1978) 'Updating defensible space', Architects' Journal, 11 October, 674

Wilson, S. (1980) 'Vandalism and defensible space in London housing estates', in Designing Out Crime, London: Home Office Research and Planning Unit

Chapter Seven

CRIME IN POLAND: TRENDS, REGIONAL PATTERNS AND NEIGHBOURHOOD AWARENESS

S.P. Bartnicki

The main purpose of this essay is to describe spatial variations of crime at state and neighbourhood levels. It will concentrate on two themes. First, the general picture of crime trends and spatial distributions of crime rates in Poland in the last decade will be presented. Second, some situational crime prevention measures taken by the inhabitants of several Warsaw neighbourhoods will be described and will be related to spatial variations of images and fears of crime.

The most recent crime wave in Poland, which reached its peak in 1984 (see Tables 7.1 and 7.2) led to a remarkable shift of attention towards the problem of crime. Both at state-official and social-individual levels various crime prevention policies and measures have been introduced. Although it is not yet possible to assess the success of these, crime rates (as measured by official statistics) are slowly declining. On the other hand social surveys clearly show that many people perceive crime rates to be rising; this is probably the reason why inhabitants of many neighbourhoods, especially in Warsaw, employ additional security devices to protect their property. There is also evidence that people are now more likely to adjust their behaviour to avoid problem areas in the city.

The first part of this chapter contains a brief discussion of the crime problem at the state level. The range of official actions designed to cope with crime is very wide. They vary from introducing (in May 1985) more severe laws for criminal offences, to the establishment of new scientific bodies such as the Research Group on Social Hazards - within the institutional framework of the Polish Academy of

Table 7.1 Number of recorded offences for four categories of crime in Poland 1980-6

	1980	1981	1982	1983	1984	1985	1986
Burglary	39239	65819	74251	85097	93158	92517	80177
Theft	63400	81039	88001	94352	97158	84848	73336
Robbery	5055	6228	6143	7277	8610	8510	7400
Rape	1576	1395	1684	1875	2184	2102	1896

Source
Rocznik Statystyczny Polski 1985, 1986 (Statistical Yearbook of Poland 1985, 1986).

Table 7.2 Crime rates in Poland 1980-6: rates per 100,000 population

	1980	1981	1982	1983	1984	1985	1986
Burglary	110.3	183.3	205.0	232.7	252.4	247.9	213.0
Theft	178.2	225.7	242.9	258.0	264.9	227.0	194.8
Robbery	14.2	17.3	17.0	19.9	23.3	22.8	19.7
Rape	4.4	3.3	4.7	5.1	5.9	5.7	5.0

Source
Rocznik Statystyczny Polski 1985, 1986 (Statistical Yearbook of Poland, 1985, 1986).

Sciences. The so-called 'May Act', intended to be in force for three years, was introduced in May 1985. This Act was passed, as one can read in its preamble, in response to the unprecedented increase in crime which had been apparent for many years. The main aim of this Act was deterrence. It was aimed both at would-be criminals who might be deterred by the penalties which could eventually be imposed and at the general public who should be aware of these penalties. This kind of 'preventive' policy has in fact as many opponents as it has adherents but does reflect current social feelings. Social polls made by the CBOS (Centre of Public Opinion Surveys) indicate that many believe in the effectiveness of severe sanctions and wish to universalize them. In fact more than 57 per cent of the sample surveyed (in the research by CBOS) think that despite the rather severe law which has been in force since June 1985, the punishment of criminals for such offences as robbery, assault, rape, theft should be even tougher and only 4 per cent think that the criminal code is too severe (Podemski 1986). What is even more striking in the results of this survey is that half of the surveyed people want changes included in the May Act to be made permanent.

On the other hand there has also been significant opposition to this kind of criminal policy (Podemski 1986). Some authors indicate that none of the stated goals were achieved. In this research, Kram and Nowakowski (1987) showed that the claim about the effectiveness of the May Act could not be confirmed from the available evidence; nor could the relationship between severity of penalties and the declining levels of crime. This research proved that the deterrent effect of the May Act could not work because offenders simply did not know about the changes in criminal code. In the sample of 295 offenders, more than 170 have never heard about this Act and many others misunderstood its regulations (Kram and Nowakowski 1987).

Although this last period of significant increase of crime is not the first one in post-war Poland it is the most clear, especially for burglary which is the most frequent and most feared crime (see Figure 7.1).

The trends for total crime and burglary in Poland for the period 1954 to 1986 are shown in Figure 7.1. Whereas burglary, the most feared type of crime, did not show a significant increase until the 1980s, the overall crime rate showed several peaks at earlier periods and generally speaking trends in Polish crime rates since 1955 can be

Figure 7.1 Crime in Poland, 1954 to 1986

Note: Rates of total crime and rates of burglary per 100,000 population were calculated.

Source:
Statistical Yearbook of Poland, 1955 to 1987.

divided into four main phases:

> Phase 1 (1954-1964) can be divided into two. From 1954 to 1958, when this wave reached its peak with 1,476 offences per 100,000 inhabitants, crime was rising rapidly. After 1958 significant decline can be observed and throughout this period rates of burglary were low.
> Phase 2 (1964-1971) was the period of high overall crime rates. The total number of recorded offences fluctuated around 500,000 annually and the rate of burglary grew slowly from 27 burglaries per 100,000 population in 1964 (the lowest rate in the whole period) to 99 per 100,000 in 1971.
> Phase 3 (1971-1980) was a period of significant decline of total crime rates, 1,449 recorded offences per 100,000 population in 1971 and only 949 per 100,000 in 1980, but at the same time witnessed the slow but constant rise of burglary rate from 99 in 1971 to 110 in 1980.
> Phase 4 (1980-1986) was the period of unprecedented rise of crime (both total and burglary) up to 1984 and slow decline after that year (see Tables 7.1 and 7.2). (This rapid rise is usually linked with economic and social instability which characterized Poland in the 1980s.)

This discussion of four phases of crime in post-war Poland serves as the point of departure for an analysis of regional distributions of criminal offences by voivodships. Although this level of geographical analysis of crime is based on very generalized statistical data it does help to draw attention to some important spatial variations and temporal trends.

The statistical information used shows the distribution of crime rates for burglary and robbery in two periods. For 1976 to 1978, a three-year average shows the distribution of crime at its lowest point in the post-war period. For 1982 to 1984 the figures reveal the period of a rapid rise of crime in Poland.

Regional variations of crime in Poland are very significant. In short, violent crime is concentrated in north-east and east Poland while property crime is concentrated in north-west voivodships. As in many other studies, the problem of crime is greatest in highly urbanized and industrialized areas. In order to obtain a statistical summary of the congruency of spatial variations between various

Table 7.3 Intercorrelations between reported offences, 1976-85

	Murder	Assault	Rape	Robbery	Burglary	Theft
Assault	0.09					
Rape	0.68	0.21				
Robbery	0.73	0.21	0.56			
Burglary	0.67	-0.33	0.43	0.61		
Theft	0.49	-0.23	0.44	0.50	0.82	
Total	2.66	-0.13	2.32	2.61	2.20	2.02

Excludes self-correlations

categories of crime, Spearman rank-order correlation coefficients were calculated.

Of interest in Table 7.3 are the very small correlations between assault and other categories of crime. The majority of coefficients are positive and show the consistency of geographical distribution of offences. The only negative coefficients in Table 7.3 are between burglary and assault (-0.33) and between theft and assault (-0.23); correlation coefficients between assault and the other categories are small. On the other hand the very high correlation coefficient between property crimes (burglary and theft) should be stressed. (All coefficients higher than 0.368 are significant at 0.01 per cent level).

The comparison of Figures 7.2 and 7.3 demonstrates the spatial consistencies of patterns of specific crime categories during the period of fast change in the general level of crime. Additional information about the temporal changes within voivodships indicates that while in general patterns of burglary and robbery vary, there are areas which are highly affected by both categories of crime. These are: Szczecin voivodship in north-east Poland (SZ), Gdansk voivodship in the north (GD), Jeleniogora voivodship in the south-west (JG), Lodz voivodship in central Poland (LD), Krakow voivodship in the south (KR) and most of all Warsaw voivodship in central Poland. All of these areas experienced the fastest increases of crime in the last decade exaggerating the disparities in crime levels. The coefficient of variation grew in the period 1977 to 1985 from 38.1 to 47.9 in the case of robbery and from 43.5 to 57.7 in the case of burglary. This means that those voivodships which were most affected by crime in the mid-1970s are relatively more affected ten years after. One should pay special attention to the case of Warsaw. The capital of Poland is placed in the highest quintile of crime rate, by voivodships, nine times.

On the other hand it should be underlined that the most urbanized and industrialized region of Silesia (Katowice voivodship (KA)) does not belong to these most affected areas and could be placed in the middle of the scale.

The neigbourhood level

The rise of crime in Poland after 1981 led to a significant growth of interest in the crime problem. This was reflected in the quantity of newspaper publications and TV and radio programmes about crime. As others have suggested (see Smith 1984) 'newspaper reporting might create and define a broad public awareness of crime that is substantially different from any "reality" contained in the official statistics' (Smith 1984, 293). It was also suggested that there are many other factors responsible for the shape of the 'surface of fear' held in one's mind: these include interpersonal gossip or rumour (Smith 1984), ecological labels (Brantingham, Brantingham, and Butcher 1982), people's position in social and geographical space (Garofalo 1981; Westover 1985). It can be also suggested that people will adjust their protective behaviour to their images of crime rather than to the 'reality'. The hypothesis tested here is that people's behaviours are to a large extent the reflections of fears rather than adjustments to the problem of crime as reflected in official statistics. All of these arguments suggest that the map of fear of crime or perceived level of crime in different areas is not necessarily congruent with the map of real crime rates. If the difference is substantial, we could make a further argument that people would adjust their protective behaviour to their perceptions rather than to reality. This argument was also raised by Brantingham, Brantingham, and Butcher (1982, 1) who argued that incongruent patterns (perceptual and objective) 'can have important objective consequences: people may wrongly avoid some places in the belief that they are dangerous places, or frequent dangerous places in the mistaken belief that they are safe'.

In Warsaw the perceived levels of crime in neighbourhoods does not match the real pattern of crime rates (Spearman's rank correlation coefficient between the two at 0.1 is not significant). Again, the correlations

Figure 7.2 Burglaries in Poland, 1976-8 and 1982-4

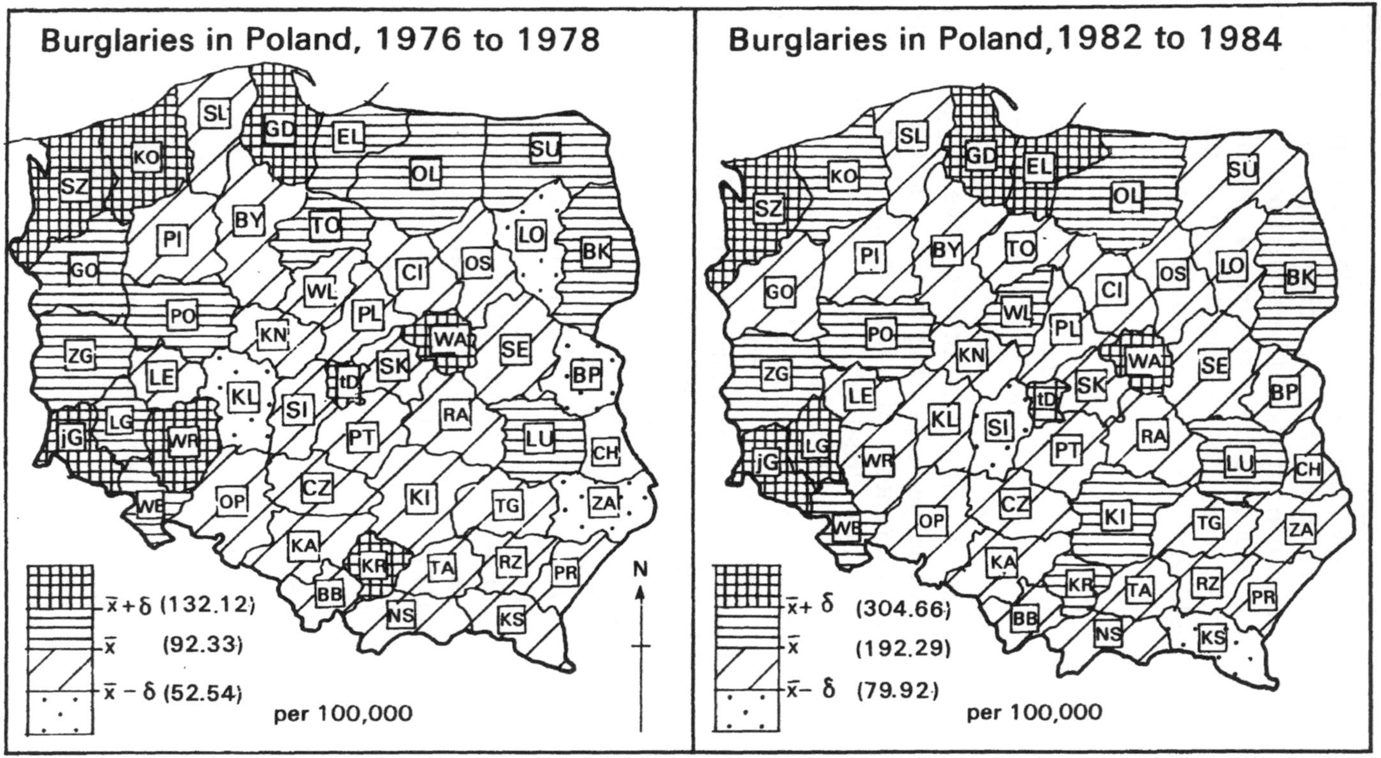

Figure 7.3 Robberies in Poland, 1976-8 and 1982-4

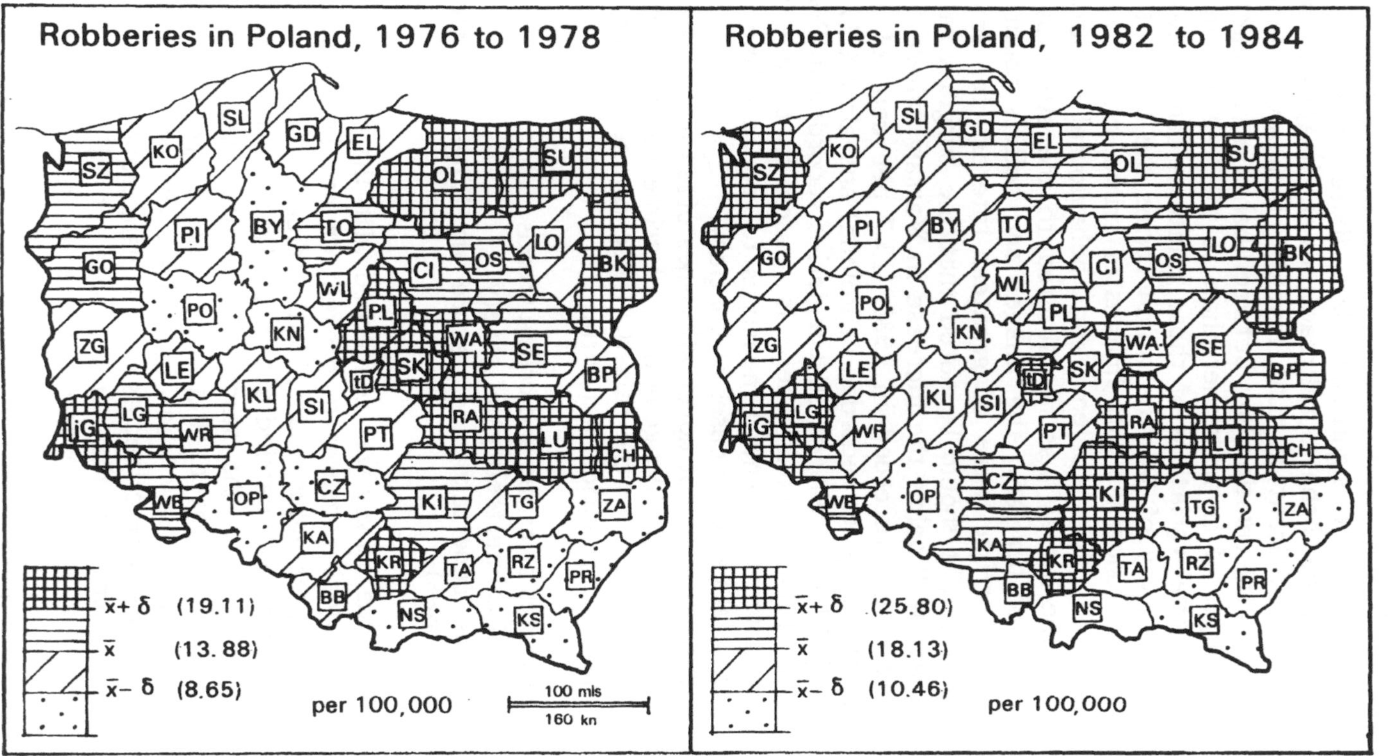

between different behavioural reactions and the 'real' levels of crime are not significant but those reflecting the relations between behavioural reactions and images of crime are often significant at 0.01 level. This analysis focuses therefore on the relationships between behavioural reactions to the images of crime rather than to the 'reality'; in this way the spatial variations of those relations can be tested. The main survey of Warsaw which forms the basis of this analysis was concerned with images of crime, crime trends, attitudes to the problem and behavioural reactions. Although the survey was not limited to the study of the perception of burglary, the images and beliefs about this category of crime were of greatest interest for several reasons. First, it should be noted that the burglary rate in Poland was well below the increase in Warsaw, where 5,395 burglaries were recorded in 1980 (13.7 per cent of total number of burglaries recorded in Poland) and almost 17,781 in 1984 (19.1 per cent) of all burglaries recorded in Poland (Bartnicki 1986). Second, there is reason to believe that situational (environmental) measures to reduce crime have high potential values in cases of burglary.

The survey did not cover all issues often undertaken by the studies of subjective geography of crime but most of the important questions were raised. These can be divided into three main groups concerned with:

(i) Estimates of crime and crime trends
(ii) Fear of crime/feelings of safety
(iii) Actions undertaken to secure dwellings (reaction - adaptation).

Generally, the survey was concerned with the perception of crime and local measures to prevent crime. The study was conducted in the autumn of 1986 (Bartnicki 1987).

Fieldwork was carried out in twelve neighbourhoods of Warsaw, chosen as the study area on the basis of earlier research which proved that strong stereotypes of crime areas exist in Warsaw (see Figure 7.4). The neighbourhoods situated in areas described as safe or dangerous were chosen on the assumption that inhabitants of different urban environments would perceive crime as a social problem differently, that they would present various attitudes towards this phenomenon and would show different levels of fear. Chosen neighbourhoods are situated in following areas: Zoliborz, Muranow, Kolo, Jelonki, Ochota Centralna,

Wlochy, Stegny, Ursynow-Natolin, Brodno, Praga Centralna, Saska Kepa, Goclaw (see Table 7.4 and Figure 7.4).

Estimates of crime and crime trends

After a five-year period of rapid crime growth in Poland (1981-5) and the accompanying massive propaganda in mass-media and official anti-crime programmes, it was of interest to find out how people estimated the problem. The results obtained were not surprising and the hypothesis presented by Brantingham, Brantingham, and Butcher (1982), that generally people tend to view crime as a problem of 'other' places was confirmed. This simply means that people believe that crime is rising in the country as a whole, is rising in their city but is not rising in their own neighbourhood. The results of the survey show that 72 per cent of respondents believed that crime was rising in Poland and just about 67 per cent thought that crime rate was rising in Warsaw.

Comparison of the answers to these questions in different neighbourhoods indicates a strong positive relationship between beliefs that crime is rising in the country and that crime is rising in Warsaw ($r = 0.91$ significant at the 0.01 level). Those respondents who felt that the trend was going up in the country also thought that crime was rising in the city. The large proportion of those who think that crime rates are growing stands in the opposition to official assurances and claims that something 'had already been done about crime'. In other words people believe that official action on the crime problem in Poland was not enough to cope with it. On the other hand the problem of crime was not something which people directly experienced, rather it was something which happens somewhere else and to somebody else. Only 20.7 per cent of respondents believed that crime is rising in their own neighbourhood. The only exception is the neighbourhood situated in Praga Centralna. Sixty per cent of respondents in this area believed that crime was rising in their own neighbourhood, an area which is in fact commonly named as the most crime-ridden area in Warsaw (see Figure 7.4). Thus the findings obtained in other studies were confirmed and it is very clear that perceptions of crime trends show a tendency for respondents to recognize a fear of crime but only at a distance.

Figure 7.4 Perception of crime in Warsaw

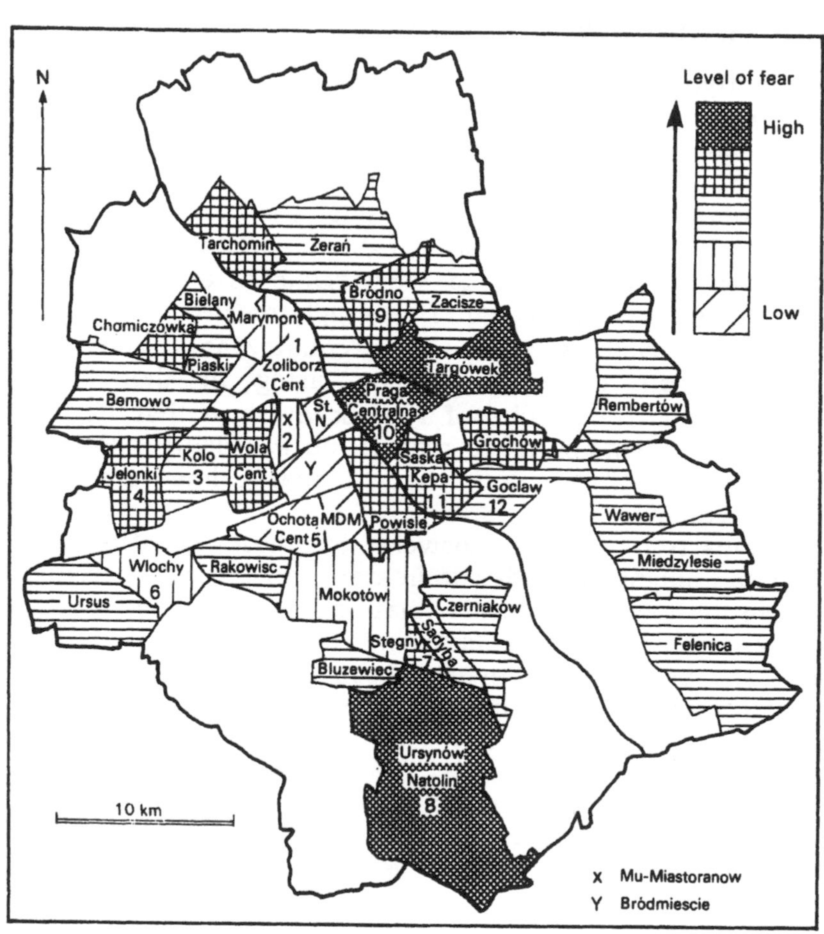

Note

The darker the shading, the worse the image; study neighbourhoods are shown. The survey was of 400 19- to 24-year olds in Warsaw.

Table 7.4 Type of housing, approximate dates of construction, and symbols used in diagrams of Warsaw neighbourhoods

	Neighbourhood	Symbol	Larger district	Date of construction	Type of housing
1	Sady Zoliborskie	SAS	Zoliborz	1966-1970	low
2	Muranow	MUR	Muranow	1949-1955	low
3	Kolo	KOL	Kolo	1945-1948	low
4	Lazurowa	LAZ	Jelonki	1976-1980	high
5	Kolonia Staszica	STA	Ochota Centralna	before 1939	low
6	Wlochy	WLD	Wlochy	before 1939	low
7	Stegny	STE	Stegny	1971-1975	high
8	Jary	JAR	Ursynow-Natolin	1976-1980	high
9	Brodno	BRO	Brodno	1971-1975	high
10	Praga	PRA	Praga Centralna	1949-1960	low
11	Saska Kepa	SAS	Saska Kepa	before 1939	low
12	Ostrobramska	OST	Goclaw	1976-1980	high

High: more than four storeys
Low: four storeys and less

Note
More than 300 respondents were surveyed and 300 questionnaires were taken into account (25 from each estate). Only people living on the ground floor were questioned. The questionnaire was partly based on the model used in a survey of Swansea (Herbert and Hyde 1984).

This was also confirmed in a question concerned with the estimation of the scale of the problem in the neighbourhood. More than 40 per cent of respondents believed that the problem of crime in their neighbourhood was small, very small, or that it was not a problem at all (19.7 per cent). Again Praga Centralna, where 60 per cent of questioned people thought that crime is a big or a very big problem in this area, proved to be the exception.

It was also assumed that there is a relationship between perceived neighbourhood crime level (the measure expressed as the proportion of those who think that crime is a big or very big problem in their neighbourhood) and the perception of existing crime trends (see Figure 7.5). It was hypothesized that those who think crime is already a problem in a given area would also believe that the problem is on the increase. Spearman's rank correlation coefficient suggests that the relationship is rather strong ($r = 0.63$ significant at 0.05 level). It is clear that people who live in areas perceived to be unsafe think that things are getting worse while those living in areas perceived as safe do not. Again there are two neighbourhoods with positions on the graph which are outstanding: Praga (PRA) and Brodno (BRO). It should be stressed at this point that existing images of crime have little in common with reality and it can therefore be assumed that inhabitants of Praga and Brodno take precautions out of proportion to the existing level of risk.

Fear of crime and feelings of safety

Although fear of crime and feelings of safety are different measures (see Brantingham, Brantingham, and Butcher 1982) they are in fact concerned with the same phenomenon. Fear of crime is linked more clearly to behavioural adjustments because people take preventive steps because of fear. On the other hand it is very important to find out whether people feel safe in certain places (especially in their places of residence) and eventually to find out what it is that makes them feel safe.

In the Warsaw survey people were asked both about fear of crime and about feelings of safety. On the latter, findings presented by Brantingham, Brantingham, and Butcher were again confirmed. Almost 50 per cent of respondents feel safer in their own neighbourhood than in other parts of the city (30 per cent 'don't know') and only 8 per cent feel less

Figure 7.5 Perceived neighbourhood crime level and the perception of crime trends in the neighbourhood

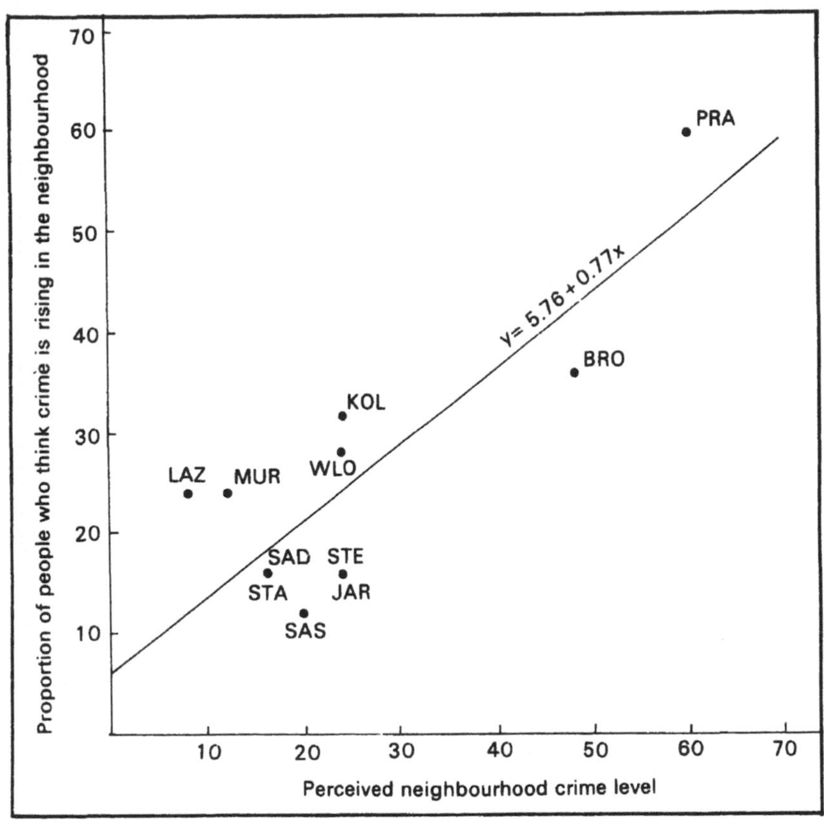

safe. It also seems that people, regardless of their fear of crime, personally feel safe.

There is a negative relation between feelings of safety in one's neighbourhood and the perceived neighbourhood crime level. The correlation coefficient between the two is -0.35. Although the relation is negative it is not strong and it is clear that without the influence of Praga and Brodno it would even be weaker (see Figure 7.6).

Figure 7.6 Feelings of safety and perceived neighbourhood crime level

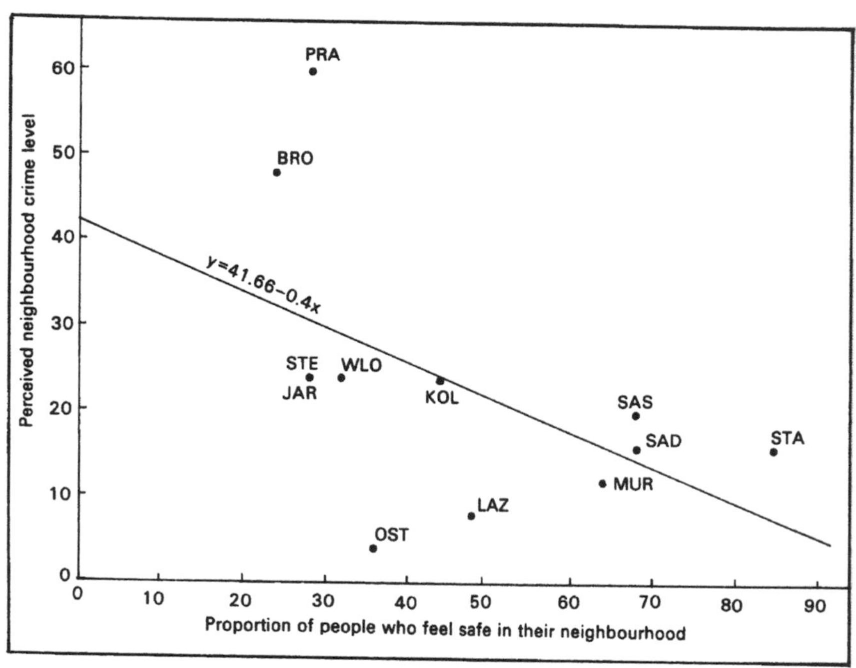

Inhabitants of both these areas perceive the problem of crime in their estates to be very big and they feel unsafe. On the other side of the graph there are the four old, small and generally desirable neighbourhoods of Saska Kepa (SAS), Sady Zoliborskie (SAD), Muranow (MUR), and Kolonia Staszica (STA) whose inhabitants feel safe in their estates and do not perceive crime as the big problem.

A very similar picture was obtained in the analysis of the relationship between the fear of crime (measured as the proportion of people who are afraid to walk at night) and the perceived neighbourhood crime level. 17.7 per cent do not go for a walk in their neighbourhood at night because of fear and 56 per cent go for a walk at night with no fear at all. Again clearly different answers were received in Praga (PRA) and Brodno (BRO). Only 24 per cent of respondents from Praga and Brodno go out for a walk at night with no fear at all and 48 per cent of respondents from Brodno and 32 per cent of respondents from Praga go out at night with some fear. Twenty-eight per cent (Brodno) and 44 per cent

Table 7.5 Number of areas named as dangerous in Warsaw neighbourhoods

SAD	KOL	MUR	LAZ	STA	WLO
11	22	14	8	16	19
STE	JAR	BRO	PRA	SAS	OST
10	16	18	21	17	10

(Praga) do not go for a walk at night because of fear. Fears exhibited by those who do not go for a walk at night in their own neighbourhood are clearly connected with the fear of violent crime (robbery, assault) but not with property crime. Although there were some behavioural adaptations in this context such as avoidance of certain areas during certain times (inhabitants of Praga neighbourhood named 21 such areas within their neighbourhood while inhabitants of Lazurowa (LAZ) only 8 and Ostrobramska (OST) named 10 areas which were avoided) (see Table 7.5), security devices linked with the fear of being burgled are much more common.

Behavioural reactions to the threat of burglary

At least two general kinds of adaptation/reaction to the fear of being the victim of burglary can be distinguished:

(i) Passive reactions such as insurance
(ii) Proactive reactions such as target hardening

Passive reactions

People insure their homes for various reasons (fire, water, wind damage, and others) but some of them insure homes because of the fear of being burgled. In surveyed areas more than 51 per cent of dwellings are insured and less than half of respondents who insured dwellings at all mentioned the fear of being burgled as the main reason for insurance. Two diagrams illustrate the relationships between levels of 'general' insurance: insurance caused by the fear of being burgled and the perceived neighbourhood burglary level. Spearman correlation coefficients are as follows: Between level of general insurance and insurance caused by fear of burglary: $r = 0.72$, significant at 0.02 level. Between general insurance and the perceived burglary level: r = 0.58 significant at 0.1 level. Between perceived level of burglary and the insurance caused by the fear of burglary: r = 0.86 significant at 0.01 level.

The low correlation between perceived burglary level and general insurance indicates that there are other factors involved. One such factor could be a particularly active insurance agency and this was mentioned as the main reason for taking insurance in one-fifth of cases. Spatially it ranged from about 40 per cent (OST, JAR, STE) to 0-4 per cent (MUR, SAS, STA).

The higher correlation coefficient between the general level of insurance and the insurance caused by the fear of being burgled suggests that fear of burglary is, in fact, a very important reason for taking insurance. This suggestion is confirmed by the very high correlation coefficient between the perceived level of burglary and the fear of burglary as the main reason for insurance (r = 0.86). Obviously the image of high vulnerability in certain areas has great significance. This finding in fact supports the assumption that it is the image of crime rather than the real level of crime which shapes people's behaviour. What is more, the correlation coefficient between the real level of burglary and the fear of burglary as the main reason for taking insurance is rather low: r = 0.16. It is again apparent that residents of these twelve Warsaw neighbourhoods react according to their images rather than to reality.

Proactive reactions

(i) Leaving the light (or radio/television) on when the flat/house is not occupied
(ii) Installing additional door locks
(iii) Protecting the dwelling by fitting iron bars on the windows

Results of the study show that all mentioned reactions to the fear of being burgled are very popular. Overall almost 55 per cent of surveyed people said 'yes' to the question 'do you leave the light (radio/TV) on when the dwelling is not occupied?' Spatially, the proportion of those who use this kind of deterrence is differentiated: the highest number of positive answers was found in eastern neighbourhoods of Ostrobramska (OST) and Praga (PRA) (76 per cent). The lowest proportion was found in western suburbs: Lazurowa (LAZ) and Wlochy (WLO) (36 and 32 per cent). It was again assumed that the proportion of those who use this kind of prevention will vary spatially according to the geographical pattern of perceived neighbourhood burg'ary level. In reality the Spearman correlation coefficient was not as high as expected (r = 0.61) significant at 0.05 level, but there is still some apparent relation.

Another proactive adaptation to the perceived neighbourhood crime level which could be seen in many places was the installation of additional door locks. Usually people who decide to fit non-standard additional door locks, install two or three of them. The correlation coefficient between the proportion of people who use additional door locks and the perceived neighbourhood burglary level is r = 0.63 (significant at 0.05 level). The average proportion of those who use additional locks is 51 per cent (very similar to the proportion of those who leave the lights on - 55 per cent). Also the spatial variation is similar (the correlation coefficient between the two is 0.74). Additional door locks are most often fitted in eastern neighbourhoods of Warsaw Praga (PRA), Ostrobramska (OST), and Brodno (BRO), where more than 70 per cent of surveyed households had locks installed.

The most visible of security devices used by the inhabitants of many neighbourhoods to protect dwellings from burglary or theft is the installation of iron bars (grills), on windows and in some cases on entrance doors. The quantity of those target-hardening devices is in some cases striking.

On the whole more than one-third of ground-floor apartments in the study areas use this kind of protection. The proportion of those who use iron bars varies from about 50 per cent in Saska Kepa (SAS) and Sady Zoliborskie to 0 per cent in Kolo (see Table 7.6).

The proportion of those who fitted iron bars is not directly related to the perceived neighbourhood burglary level (r = 0.22), nor to the spatial distribution of other kinds of protective activities. The correlation coefficient with the proportions using additional door locks is only -0.07 with those who are insured (r = 0.08), and with the proportion of those who leave the lights on when nobody is at home (r = 0.18). It seems that there are other reasons than the fear of being burgled which lead people to fit bars. Surveyed people were asked about these reasons. Although more than 50 per

Plate 1 Barricaded ground-floor windows in Warsaw

Table 7.6 Proportion of those who use iron bars on ground-floor windows in twelve Warsaw neighbourhoods

SAS	KOL	MUR	LAZ	STA	WLO
48	0	24	20	36	44
STE	JAR	BRO	PRA	SAS	OST
24	28	28	24	48	40

cent confirmed that the reason was the fear of burglary (half of those 50 per cent mentioned that they fitted bars after they experienced burglary), 31 per cent said that it was the suggestion of the agent of the insurance company and more than 5 per cent that the reason was the example of neighbours. This last reason seems to be even more important. There are many blocks of flats in surveyed areas where all ground-floor windows are barricaded (see Plate 1).

Sometimes it is not the 'example' of the neighbours but rather the fear that neighbour's iron bars can be used as the ladder. This leads to fitting iron bars on the first floor (Plate 2) or even on the second floor, especially when there are bars below.

Quite often iron bars are installed on the top floor and here the easy access from the roof is the reason (Plate 3). Very often the design of the top floor creates obvious situational opportunity for would-be burglars, and inhabitants react by fitting iron bars to change situational opportunities for entering the dwelling.

Sometimes bars are fitted in unexpected places in the middle of blocks where inhabitants are afraid that the burglars would use balconies as a way of reaching the flat.

Apart from more or less obvious places to fit the iron bars people sometimes install them inside the building. They protect individual entrance doors or the whole interior corridor. A staircase, for example, can be separated from about five dwellings by a specially designed door and each dwelling unit can be contacted by the visitors so that the door would then be opened directly from the dwelling (electric lock).

Two further points help to explain the low correlation between the incidence of bars and both perception of being burgled and the distribution of other security measures:

First, local authority regulations (fire regulation) can be enforced in different areas with different strength. Again, iron bars are rather expensive target-hardening

Plate 2 Barricaded first-floor windows in Warsaw

Plate 3 Barricaded upper-floor windows in Warsaw

devices and many people cannot afford to fit them. The suggestion by Clarke (1983) that 'for some people situational prevention has unattractive connotations of ... a "fortressed society" ...' can be to some extent supported (at least in the case of those who employ situational measures (bars)). More than 45 per cent of those who fitted bars said that they were not happy with the idea of 'fortressing' their dwellings. On the other hand a large majority of those who installed iron bars (91 per cent) said that after fitting them they felt much more secure.

To summarize on the theme of target hardening it can be said that the obvious desire of people to make their dwellings more secure finds its expression in space. People change situations perceived as vulnerable using different forms of protection. The most popular are additional locks and iron bars. More sophisticated devices are not yet common. It is clear that people know exactly which places are most vulnerable and why. This statement can be supported by the fact that only just under 40 per cent of bars in the study areas were fitted on corner locations which are usually perceived as most vulnerable. From other studies we know that corner locations are in fact those which suffer the most (Herbert and Hyde 1984).

Summary

It is clear that although the problem of crime in Poland may not be as serious as in many western countries it does attract considerable public attention.

As in other countries, crime rates in Poland vary significantly over space. Although generally it can be said that the regional variation of crime rates is stable in time, the increase in the period 1980-85 was not uniform in all parts of the country. In some areas crime rates did not rise at all. In others the growth was substantial. During this time regional variation deepened. The most affected areas were the highly urbanized, with the exception of the Silesian region. The highest crime rates are in Warsaw, where the rise, especially of burglary, was well above the national average.

Recent increases in crime have caused both official and public reactions though of a very different kind. At the state level, anti-crime policies were implemented with the aim of deterring would-be offenders by more punitive sanctions. The implementation of deterrence sentencing is officially believed to be one of the main reasons for the slight decline of crime rates in 1986 and 1987. A policy of greater severity of punishment for most criminal offences is supported by a large proportion of society and this support is reinforced by a general belief that despite official claims to the contrary, crime rates are still rising.

At the neighbourhood level various kinds of situational measures are being applied. The widespread nature of these measures reflects the fact that fears of victimization far exceed the actual levels of risk. Indeed, fear of crime is a much more serious problem than that of the crime itself and it is to fears rather than to realities that people react. Levels of fear and levels of implementation of different situational measures are highly congruent. The most popular preventive measures are additional door locks and iron bars; other measures include leaving on the light (radio/TV) when the dwelling is not occupied.

The spatial variation of target-hardening measures in Warsaw is related to the perceived levels of risks as were other behavioural adaptations such as avoidance of specific areas thought to be unsafe. It is therefore possible to conclude that the research into the geography of crime in Poland is based as much on the analysis of subjective as of objective levels of crime. Fear of crime is easily induced and has considerable consequences for the quality of life in those urban neighbourhoods where it is prevalent.

References

Bartnicki, S.P. (1986) 'The geography of crime: a case study of Warsaw', in Miscellanea Geographica, Warsaw: Faculty of Geography and Regional Studies, 237-43

Bartnicki, S.P. (1987) 'Ecological labelling and the patterns of crime in Warsaw', paper presented at the conference Solutions to Pathologies of Urban Processes, Kazimierz Dolny, 12-16, October 1987

Brantingham, P.L., Brantingham, P.J. and Butcher, D. (1982) 'Perceived and actual crime risks: an analysis of inconsistencies', 21st annual meeting, Western Regional Science Association, Santa Barbara, Feb. 1982

Clarke, R.V.G. (1983) 'Situational crime prevention: its theoretical basis and practical scope', in Tonry, M. and Morris, N. (eds) Crime and Justice: An Annual Review of Research, Chicago: The University of Chicago Press

Garofalo, J. (1981) 'The fear of crime: causes and consequences', Journal of Criminal Law and Criminology 72, 839-57

Herbert, D.T. and Hyde, S. (1984) Residential Crime and Urban Environment, Swansea: Department of Geography, University College of Swansea

Kram, A. and Nowakowski, W. (1987) 'Stop dla Ustawy Majowej' (Stop for the May Act), in Prawo i Zycie, Law and Life, 39, 3-4

Podemski, S. (1986) 'Co myslimy o praworzadnosci' (What do we think about obeying the law?), Polityka 32, XXX, 3

Smith, S.J. (1984) 'Crime in the news', British Journal of Criminology, 24, 3, 289-95

Westover, T.N. (1985) 'Perceptions of crime and safety in three midwestern parks', Professional Geographer 37(4), 410-20

Chapter Eight

BEHAVIOURAL GEOGRAPHY AND CRIMINAL BEHAVIOUR

George Rengert

The objective of this chapter is to examine the relevance of concepts developed in behavioural geography to the study of criminal behaviour. The intent is to illustrate that many of these concepts could potentially occupy the cutting edge of ecological criminology. Some of these concepts have not yet been applied to criminological problems but hold promise of explaining relationships which are as yet unclear. Other geographic concepts may suggest practices useful to criminal justice practitioners. Before discussing the relevance of behavioural concepts to criminal justice problems, consider the place of behavioural geography within the discipline of geography.

Behavioural geography is implicit in much of human geography. It explicitly became a focus of geographic interest in the 1960s when researchers realized that neoclassical economic assumptions borrowed for use in location theory deviated too far from reality (Cox and Golledge 1981). For example, gravity models were used to explain spatial patterns which were outcomes of individualistic decentralized decision-making processes. In these models, perfectly informed utility-maximizing humans made locational adjustments in response to the spatial disequilibria in rewards emanating from their actions. The resulting locational adjustments in turn re-established equilibrium. This model of spatial activity assumed 'economic man' who had perfect knowledge of the alternatives available and used this knowledge in an economically rational manner. Behavioural geography emerged with the recognition that such assumptions had little in common with real-world relationships and left a great deal to be desired (Wolpert 1964).

Behavioural geographers identified and evaluated important questions concerning imperfect knowledge on the part of location decision-makers. Such questions included how many locational alternatives individual decision-makers have knowledge of, and which of these alternatives are actually considered in a locational decision process. Concepts such as mental maps, awareness space and decision theory were adopted by geographers in an attempt more closely to mirror reality.

Geographers working in the field of criminology began to adopt these concepts to explain criminal spatial behaviour which often did not make sense from a strictly economic standpoint (Carter and Hill 1979; Rengert and Wasilchick 1985). However, before these weaknesses of the gravity model were fully realized, several researchers had applied them to criminological problems (Smith 1976; Felson 1986; Rengert 1981). The results were predictable. Not only were incorrect assumptions implied concerning the criminal's knowledge of space, but, more importantly, incorrect conclusions resulted from a comparison of actual patterns with the hypothetical ideal derived from the gravity model.

Several components of behavioural geography are useful for understanding criminal behaviour, and we will briefly review these components and how they can be applied to problems in criminology and criminal justice. First, let us consider the broadest categories into which research in behavioural geography has been classified. The primary focus of behavioural geography is the spatial decision making of individuals which precedes their behaviour. In the broadest sense, these decisions can be analysed as (i) spatial behaviour, and (ii) behaviour in space.

Research in spatial behaviour consists of a search for models and concepts to describe behavioural processes irrespective of the spatial structure in which the behaviours are found. When mapped, criminological theory which focuses on the location of criminals is in this vein. Such theories attempt to explain why a person located in a specific social, economic or physical environment decides to commit a criminal act. For example, Merton's theory of anomie suggests that a low-income minority population may contain a disproportionate number of criminals due to barriers placed in their path toward attaining the legitimate goals which are defined by the larger society. Frustrations lead to criminal acts as a means of attaining these goals by illegitimate means (Merton 1987). Therefore, the spatial arrangement of poor minority people who are expected to experience anomie defines the spatial arrangement of crime rates.

Most distance and directional studies in the geography of crime are also in this vein. Capone and Nichols (1976) compared the distance travelled from home to crime site for robbers in Miami, Florida. Costanzo, Halperin, and Gale (1986) studied the direction criminals are likely to travel from their home to commit a crime. Neither of these studies considered the spatial structure of the environment within which the criminals were interacting.

Behaviour in space relates actual human behaviour in an area to specific spatial structures of that region. Opportunity theory in criminology, in which behaviour in space is constrained by the relative location of opportunities for crime, is one example (Mayhew et al. 1976). It presumes that the spatial structure of an area as manifested in the relative location of opportunities for crime can be used to explain criminal behaviour in space.

In the beginning, behavioural geography developed around concepts of spatial behaviour rather than behaviour in space. This was an important occurrence since it emphasized the individual as an active agent in the environment rather than a simple reactor to physical and social elements within this environment. Thus, the individual was viewed as an active decision-maker rather than simply as a reactor to the spatial structure of the environment. Individuals were assumed to process information about their environment through different perceptual filters. Decision-making, choice theory, analysis of space preference, spatial learning processes, environmental cognition and cognitive mapping have all developed in geography out of the process of identifying the active decision-making component of human behaviour. These concepts have obvious potential for expanding criminological theory.

Outside of North America, behavioural research in geography was more likely to focus on behaviour in space. This thrust was spearheaded by the work of Hagerstrand (1970) in Sweden and also characterized some of the behavioural work in geography undertaken by British geographers (Thrift 1977). The general model is of humans as reactive decision-makers subject to a number of socio-spatial constraints. In other words, this approach is much more concerned with explaining human behaviour in terms of societal, environmental and institutional constraints on spatial behaviour. To this author's knowledge, few of the concepts developed from this perspective have been directly applied in criminological research. This approach holds tremendous promise for explaining not only criminal behaviour, but also correctional practices designed to control crime by placing exactly these constraints on the behaviour of convicted criminals.

In the following section, concepts and postulates developed by these two groups of researchers will be evaluated in terms of their usefulness in criminological research. By usefulness we mean that they help us to understand criminal spatial phenomena, to identify spatial consistencies in criminal behaviour, and to predict future spatial patterns in crime.

Behavioural geography in criminology

The behavioural ideas outlined above are as implicit in criminological theory as they are in human geography. Even in the criminological research of the nineteenth century (Guerry 1831) one can find explicit consideration of behavioural concepts of a geographic nature. However, widespread acceptance of the behavioural approach to spatial problems in criminology has never occurred. Although examples of this research are scattered throughout the geographic and criminological literature, recognition of their importance is new. In North America several scholars now have advanced this focus to a point where it is at the cutting edge of ecological criminology (Cohen and Felson 1979). It is time to take stock of the accomplishments to date.

Criminological research of spatial behaviour

Behavioural geographers identified an important omission in criminological research when they criticized studies which did not include relative location as an independent variable. Much as Hagerstrand (1970) asked 'What about people in regional science?', behavioural geographers asked, 'What about people in the geography of crime?' Knowing where criminals are concentrated did not tell how criminals used the space available to them, how much space a typical criminal actually had knowledge of, or whether the physical characteristics of a house and/or neighbourhood would even be considered in a criminal's choice process. In other words, we needed to know the 'awareness space' of criminals and how it varies with characteristics of individuals and by their relative location in space.

Behavioural geographers have discovered regularities in spatial awareness which varies with characteristics of individuals (Amedeo and Golledge 1975). Some of this work has focused on criminals. For example, Rengert and Monk (1982) have shown that the extent of spatial knowledge of female criminals is less than that of male criminals. Rengert (1975) has demonstrated that this difference affects crime patterns of males and females even when a single type of crime (such as residential burglary) is practised by both. Female burglars commit crimes closer to home than male burglars. When a female burglar does travel an extended distance, she is more likely to travel toward the central city shopping district while males operate in more scattered locations.

Carter and Hill (1979) demonstrated that ethnic characteristics are associated with differences in the spatial awareness of criminals. Black criminals were not likely to be familiar with white areas and white criminals were not very familiar with black residential areas. Rengert and Wasilchick (1985) discovered that black burglars had less spatial knowledge than the white burglars analysed in their study.

The residential locations of the criminals also impacts the extent of their spatial knowledge. Inner-city youth know less extended space but are very familiar with their own neighbourhood. Suburban youth know more extended space but have less familiarity with any one given location (Orleans 1968). Much of this difference is related to the mode of transportation used in each community. Most movement in a very congested inner-city area is by public transportation or by foot. Walking allows one to gain intimate knowledge of the surrounding neighbourhoods. The direction in which one generally walks will skew the shape of this very well known space.

The shape of known space is generally skewed in the direction of the central business district for inner-city residents. The central business district (or a regional shopping strip) is the second most used area after the home neighbourhood. Therefore, the locations of crime sites also are expected to be skewed in the direction of the central business district for inner-city criminals. Lentz (1986) demonstrated that this directional bias can change over time with changing transportation technology and changes in the spatial structure of urban areas. With the development of suburban shopping malls and employment centres, robbery trips shift somewhat towards a suburban destination. Rengert and Wasilchick (1985) demonstrated that the crime trips of suburban burglars are more likely to be skewed in the direction of other nodal use centres such as workplaces or recreation sites. Rengert (1987) illustrated the attraction of Atlantic City gambling for criminals who victimize the surrounding community.

Finally, we do not know the relationship between extent of awareness space and the type of crime or the mix of crimes a person is likely to commit. Certainly, a constricted knowledge of space limits the opportunities for crime open to an individual. A key question is whether the decision to commit a new type of crime is likely to encourage an individual to explore for new crime sites or whether knowledge of these new crime sites obtained through passive travel leads to their exploitation. For example, armed robbers are not likely to operate within their home communities where they could be readily identified. Rengert and Bost (1987) demonstrated that robbers living in housing projects exhibit behaviour designed to avoid recognition by travelling further from home to commit their crime than burglars. Therefore, it is possible that armed robbers are individuals with extensive spatial knowledge who are used to being in more distant places. Capone and Nichols (1976) also demonstrated that armed robbers travel further on average to commit their crimes than burglars. It remains for behavioural geographers to identify what other decisions and behavioural processes actually make up the human activity of armed robbery, as well as other crimes.

Indeed, there is very little understanding of how criminals learn about space other than through their daily routine activities. It is taken for granted that much crime in low-income communities occurs because they contain individuals with high incentives for crime who go about their daily routines keeping an eye peeled for opportunities for crime. If they spot an unlocked car door or valuable property left on a front porch, they immediately take advantage of this opportunity for crime. This obviously is a relatively unplanned criminal activity. Unplanned crime generally is contained well within the 'awareness space' of the criminal, such as on a well travelled route or within a well used area. Therefore, public and private policies which extend the routine activity routes of a population with high criminal tendencies are seen to impact communities where this extension occurs. For example, owners of some suburban shopping malls have resisted the extension of public transportation to their property, believing that if the malls are kept inaccessible to low-income individuals without access to an automobile, they can be kept relatively crime-free. This is what Gold (1970) refers to as a 'sanitized corridor' when he speaks of expressways as a means of sifting out low-income people from the destination of the expressway. On the other hand, large sports complexes and recreational centres can attract individuals into an area they would not otherwise know. These facilities have a negative impact on their immediate neighbours which has rarely been investigated by geographers (for an exception, see Hakim and Rengert 1986).

The extent of 'awareness space' does not necessarily constrict future criminal activity since the amount of space that a criminal has knowledge of can always be expanded. If criminals practise 'spatial exploration' by travelling into unknown territory to locate crime sites, the extent of their 'awareness space' at any time has little importance. However, contrary to the popular belief of the public and some law-enforcement personnel, recent research (Reppetto 1974) illustrates that such travel is relatively rare. Most criminals not only like to operate in areas they have some knowledge of but actually prefer areas they are very familiar with (Brantingham and Brantingham 1984). Therefore, the extent of spatial knowledge and familiarity are important considerations in understanding the probable location of a crime site.

Much more research is needed to discover how spatial awareness varies by characteristics of the criminal. For example, spatial awareness should vary with age of the criminal although this has never been verified. There also may be systematic differences between individuals of different economic and educational levels. Finally, certain life experiences are likely to be important. Service in the armed forces certainly opens the spatial horizons of many young men and women. In the United States, bussing of school children into neighbouring school districts can do the same for youths. We know the positive impacts of many of these practices; the implications for criminal behaviour have not been investigated.

Another issue for investigation by behavioural geographers is how a criminal's awareness space is evaluated for criminal purposes. In other words, what criteria are used by the criminal to identify a subspace within their total awareness space which they evaluate for criminal purposes. This subspace may be defined as the criminal's 'search space' - another concept borrowed from behavioural geography (Gould 1965). It is instructive to examine how this concept has been used in geography.

Geographic research has focused on the process by which a potential purchaser identifies a new house to buy in the real estate marketplace, yet little attention has focused on how residential burglars identify homes to victimize (for exceptions, see Carter and Hill 1979; Rengert and Wasilchick 1985). Geographic research also has focused on how individuals choose which stores to patronize, but little research has focused on how shoplifters choose which stores to steal from or how armed robbers choose which stores to rob (an exception is Duffala 1976). These relatively neglected areas could make effective use of several important geographic and economic concepts. For example, are criminals of a specific type more likely to be economic satisficers and operate closer to home rather than to be optimizers who choose the best opportunity no matter how distant? Are intervening opportunities important considerations as criminals choose the nearest rather than the 'best' opportunity? These questions require careful consideration before public policy concerning the location of public facilities, transportation routes, and even correctional practices in criminal justice can be formulated carefully.

Most of the research in spatial behaviour of criminals is based on data of 'revealed space'. In other words, research has focused on the actual use of space by criminals (such as which crime sites they actually choose) and then behavioural postulates are ascribed to these spatial patterns. Geographers who have attempted to go beyond revealed space to understand how spatial decisions are actually formulated by criminals are quite few (Ley 1974). This is a research focus which holds considerable promise. As advocated by Ley (1981), it might best be practised using a humanistic approach where the researcher allows the criminal to act as a source of information and ideas rather than approaching the research with preconceived ideas of what categories of information are to be gathered.

Criminal behaviour in space

Criminal behaviour in space is a geographic approach to the study of crime which explicitly considers the geographic distribution of opportunities for crimes and the social, economic, physical, and physiological constraints on criminal spatial behaviour. This research is place-specific to the extent that the particular mix of opportunities and constraints identifies a specific location. However, categories of constraints can be used to compare criminal spatial behaviour in several areas. Also, the spatial structure of opportunities for crime can explain crime patterns which are otherwise obscure.

Criminologists have long identified opportunities for crime as important variables in explaining the spatial distribution of crime. For example, the commonly identified 'urban crime gradient' which illustrates that crime rates tend to decrease as distances from the centre of the city increase has been associated with the spatial arrangement of opportunities for crime. This point is illustrated by the fact that most white-collar crime takes place in the central business district where most banks are headquartered, corporate centres are located, and major service industries are centred. Shoplifting is higher in the central business district because large department stores are located there. In other words, opportunity theory might be illustrated by the words of Willie Sutton when asked why he robbed banks: 'because that is where the money is'. Obviously, the spatial pattern of crime is going to be determined in part by the spatial pattern of opportunity for crime. Geographers may consider this fact uninteresting, if not trivial. The type and distribution of barriers which make opportunities unapproachable from a spatial perspective may be more interesting, and also more useful from a policy perspective.

Social barriers to the attainment of an opportunity are termed constraints in the geographic literature (Pred 1981), Hagerstrand (1970) categorizes these constraints into three types: (i) authority constraints, (ii) capability constraints, and (iii) coupling constraints. Although these concepts have not been directly applied to criminology and criminal justice research, each holds considerable promise of enlightening clouded relationships. Correctional practices which are designed to ensure that criminals do not repeat old criminal practices can be categorized using these concepts. These ideas will become clear as we discuss each constraint in turn.

The most obvious constraint used in criminal justice is the 'authority constraint' in the form of law. Authority constraints may be defined as those general rules, laws, economic barriers, and power relationships which determine who does or does not have access to specific domains at specific times to do specific things. In fact, in the Marxist literature, laws are defined as power relationships (Bohm 1982). The geography of power relationships forms the basis of the work of Lowman (1986), a geographer working on criminal justice issues. Lowman argues that crime cannot be interpreted without an understanding of the power relationships which not only determine laws, but which also determine who receives the sanctions of these laws.

Much of the correctional philosophy in criminal justice is composed of authority constraints with geographic properties. For example, imprisonment is nothing more than an authority constraint on spatial movement (one cannot leave the spatial jurisdiction without permission) and on time use (one must be gainfully employed). Curfew is an authority constraint in both space and time use - a person must be in a specific place at a specific time. The expanding use of electronic tracking and monitoring devices also is largely an authority constraint with geographic properties, constraint on where a convict can be at a certain time.

These topics are largely unexplored by geographers. Yet they are obviously issues where geographic perspectives could add to their understanding. It is surprising that the topics are largely untouched. In fact, few geographers have ventured into the study of corrections. They have concentrated on the study of crime with a scattering of studies of law. Behavioural geographers have much to offer in correctional research - an area in need of careful evaluation and analysis.

The second constraint is the 'capability constraint'. It is defined as those activities which individuals can undertake because they are physically capable. It is largely a time constraint. An individual's activity choices may be limited because of the physiological necessity of allocating large amounts of time to sleeping, eating, and personal care.

Also, transportation technology limits the distance they are physically able to travel in a given time period. In other words, it asks the question of what is possible given a certain number of hours in a day and the non-discretionary tasks that must be performed at a certain place at a certain time of this day.

Some excellent geographic work has focused on capability constraints in a de facto manner. Janice Madden (1977) illustrates that contrary to accepted economic wisdom which states that employers who discriminate in wage rates lose out in the long run by receiving second-rate talent, it is profitable for an employer to discriminate against married women who are tied to a constricted commuting range by the capability constraint of time available to commute. This time is reduced even further if small children are contained in her family. Madden illustrates why formal laws are necessary where classic economic principles do not necessarily lead to a fair society. Since the Thatcher and Reagan governments in Great Britain and the United States, with their emphasis on free market principles, this type of geographic study has become very important in identifying inequalities within the economic infrastructure where government regulations may be justified and demanded.

The third constraint is the 'coupling constraint'. Coupling constraints identify where, when, and for how long an individual must join other individuals (or objects) in order to perform an activity. In the criminal justice literature, opportunity theory deals with matters of coupling constraints. For example, one cannot embezzle a bank easily unless one is a bank employee with access to the necessary infrastructure. In fact, much white-collar crime which goes undetected and unreported requires access to the organizational structure of large corporations and businesses. Those without access experience coupling constraints which prevent the completion of this type of criminal activity.

Recently the concept of 'routine activity' has been used in crime research. Following this approach, crime is dependent on a motivated criminal coming into contact (in space and time) with a victim without the benefit of guardianship. (A house guardian might be a non-employed spouse.) The routine activity concept is nothing short of coupling constraints. In other words, if one element is removed (or added in the case of a guardian), the crime is not likely to take place due to a coupling constraint. The entire idea behind residential burglary is to identify a residence which is unoccupied. The lack of a guardian provides an opportunity for a burglar to break into a house undetected. On the other hand, the presence of a guardian provides a coupling constraint since the burglar cannot come into contact with an unoccupied house (at least at this time and place).

Those who commit crimes in groups of two or more individuals depend on the assembly of the group in time and space for criminal activity. Gang violence, illegal prostitution (which depends on a customer identifying a prostitute in time and space), and the sale of illegal drugs which depends on a customer locating a source of supply in the form of a sales person are all activities subject to coupling constraints.

Let us focus on the possible use of coupling constraints in corrections. One of the major determinants of crime is the employment status of a potential criminal (Ross 1977). However, we do not know the reason for the relationship between unemployment and crime. It is generally assumed that unemployment creates economic incentives which can drive a person to crime. However, Rengert and Wasilchick (1985) identified many residential burglars who had quit jobs because they interfered with the time required for the criminal activity which they enjoyed more than their legitimate jobs. These authors suggest that research designs which incorporate questions to determine whether a criminal was employed while engaging in criminal activity do not go far enough. The researchers should probe to determine why the criminal was unemployed. Criminal activity may be a conscious choice (Cornish and Clarke 1986).

Correctional policy can use coupling constraints to keep potential criminals from the time and place necessary to complete criminal activity. For example, if homes are most likely to be unguarded during the workday (nine to five o'clock), a probation or parole agreement might require that a convicted burglar be employed during these times. The potential burglar cannot be in two places at the same time. Also, probation and parole agreements which require that the convict not knowingly fraternize with other convicted felons is a coupling constraint, although not a reasonable one in some inner-city communities where as high as one in four males above the age of twenty have criminal records.

Requirements that persons convicted of drunken driving do not enter an establishment selling alcoholic beverages is a coupling constraint. So, in a way, is the requirement that they attend counselling sessions to aid them in breaking their dependency on alcohol. Correctional practices designed to keep an individual from coming into contact with an incentive or opportunity for criminal activity can be classified as a coupling constraint. Coupling constraints differ from authority constraints since the latter are designed to control time and space use in general, not just to specific locations at specific times where criminal incentives and opportunities may be present.

Conclusions

Although it may have reached its peak in the discipline of geography, spatial behavioural research in criminal justice is still in its infancy. Research into correctional practices and research into spatial ideas embedded in the concept of the 'reasoning criminal' are especially needed. In the recent past, geographic work in criminal justice has depended too heavily on aggregate data that are easily obtainable from government sources. Relatively rare are geographic surveys of individual criminals designed to uncover information on how they make spatial decisions. Geographers have important contributions to make. Especially important, I predict, are the impacts this work will have in the correctional philosophy and practices in use today.

References

Amedeo, D. and Golledge, R. (1975) An Introduction to Scientific Reasoning in Geography, New York: Wiley

Bohm, R. (1982) 'Radical criminology: an explanation', Criminology, 19, 565-89

Brantingham, P.J. and Brantingham, P.L. (1984) Patterns in Crime, New York: Macmillan

Capone, D. and Nichols, W. (1976) 'Urban structure and criminal mobility', American Behavioural Scientist 20, 199-213

Carter, R. and Hill, K.Q. (1979) The Criminals' Image of the City, New York: Pergamon

Cohen, L. and Felson, M. (1979) 'Social change and crime

rate trends: a routine activity approach', American Sociological Review 44, 588-608

Cornish, D. and Clarke, R.V. (eds) (1986) The Reasoning Criminal: Rational Choice Perspectives on Offending, New York: Springer-Verlag

Costanzo, C., Halperin, W., and Gale, N. (1986) 'Criminal mobility and the directional component in journeys to crime', in R. Figlio, S. Hakim, and G. Rengert, Metropolitan Crime Patterns, New York: Willowtree

Cox, K. and Golledge, R. (1981) Behavioural Problems in Geography Revisited, New York: Methuen

Duffala, D. (1976) 'Convenience stores, armed robbery, and physical environmental features', American Behavioural Scientist 20, 227-46

Felson, M. (1986) 'Predicting crime potential at any point on the city map', in R. Figlio, S. Hakim, and G. Rengert, Metropolitan Crime Patterns, New York: Willowtree

Friel, L. and Vaughn, J. (1986) 'A consumer's guide to the electronic monitoring of probationers', Federal Probation 50, 3-14

Gold, R. (1970) 'Urban violence and contemporary defensive cities', Journal of the American Institute of Planners 36, 146-59

Gould, P. (1965) 'A bibliography of space searching procedures', Research Notes, Department of Geography, University of Pennsylvania

Guerry, A. (1831) Essai sur la Statistique Morale de la France, Paris: Crochard

Hagerstrand, T. (1970) 'What about people in regional science?' Papers of the Regional Science Association 24, 7-21

Hakim, S. and Rengert, G. (1986) The Natural Rate of Crime: Short- Versus Long-run Effects of Policing, paper presented to the American Society of Criminology, Atlanta

Lentz, R. (1986) 'Geographical and temporal changes among robberies in Milwaukee', in R. Figlio, S. Hakim, and G. Rengert, Metropolitan Crime Patterns, New York: Willowtree

Ley, D. (1974) The Black Inner City as Frontier Outpost, Washington: Association of American Geographers

Ley, D. (1981) 'Behavioural geography and the philosophies of meaning', in K. Cox and R. Golledge (eds) Behavioural Problems in Geography Revisited, New York: Methuen

Lowman, J. (1986) 'Conceptual issues in the geography of crime: towards a geography of social control', Annals, Association of American Geographers 76, 81-94

Madden, J. (1977) 'A spatial theory of sex discrimination', Journal of Regional Science 17, 369-80

Mayhew, P., Clarke, R., Sturman, V., and Hough, M.J. (1976) Crime Opportunity, Home Office Research Study, 34, London: HMSO

Merton, R. (1957) Social Theory and Social Structure, Chicago: Free Press

Orleans, P. (1968) Differential Cognition of Urban Residents' Effects on Social Scale of Mapping, paper presented to American Sociological Association, Boston

Pred, A. (1981) 'Of paths and projects: individual behaviour and its societal context', in K. Cox and R. Golledge (eds) Behavioural Problems in Geography Revisited, Methuen, New York

Rengert, G. (1975) 'Some effects of being female on criminal spatial behaviour', Pennsylvania Geographer 13, 10-18

Rengert, G. (1981) 'Burglary in Philadelphia: a critique on an opportunity structure model', in Brantingham, P.J. and Brantingham, P.L. (eds) Environmental Criminology, Beverly Hills: Sage

Rengert, G. (1987) The Location of Public Facilities and Crime, paper presented to Academy of Criminal Justice Sciences, St. Louis

Rengert, G. and Monk, J. (1982) 'Women in crime', in A. Rengert and J. Monk (eds) Women and Social Change, New York: Kendall Hunt, 7-10

Rengert, G. and Wasilchick, J. (1985) Suburban Burglary: a Time and a Place for Everything, Springfield, Thomas

Rengert, G. and Bost, R. (1987) The Spillover of Crime from a Housing Project, paper presented to Academy of Criminal Justice Sciences, St. Louis

Reppetto, T. (1974) Residential Crime, Cambridge, Mass.: Ballinger

Ross, M. (1977) Economic Opportunity and Crime, Renouf, Montreal

Smith, T. (1976) 'Inverse distance variations for the flow of crime in urban areas', Social Forces 54, 804-15

Thrift, N. (1977) 'Time and theory in geography', Progress in Human Geography 1, 65-101

Wolpert, J. (1964) 'The decision process in a spatial context', Annals, Association of American Geographers 54, 537-58

Chapter Nine

BURGLARS' CHOICE OF TARGETS

Trevor Bennett

Introduction

The research on burglars' perceptions and decision-making, on which this chapter is based, arose out of an interest shared by a number of academics during the late 1970s and early 1980s in the situational determinants of offending. The research was influenced in particular by the work of Ron Clarke (1) and other researchers at the Home Office Research and Planning Unit (HORPU) who had been involved in studying the importance of opportunities in the creation of crime. (2) The perspective underlying these and other opportunity-oriented studies became known as the situational approach.

Briefly, the situational approach placed a new emphasis on the role of situational factors in the commission of crime. It was believed that earlier dispositional approaches, which focused on the development of a general disposition to offend, were unable to explain a great deal of everyday crime. It seemed plausible, for example, that many offenders, such as shoplifters, committed crimes infrequently and terminated their offending after perhaps only a short period of time. The decision to offend might, therefore, be better explained not so much by an acquired or inherited disposition, but by the constraints or inducements associated with particular situations. An important weakness of the dispositional approach, which was not shared by the situational approach, was its failure to explain the commission of specific types of offence and, in particular, its failure to explain the relationship between offender and victim.

The theoretical basis of this movement was not novel. The importance of situational factors in the creation of crime had been noted by criminologists, sociologists, and psychologists for some time. Nevertheless, the approach consolidated a line of thought which until then had only been loosely articulated and intermittently explored.

An important assumption underlying the situational approach is that potential offenders are to some extent rational in their choice of targets. A useful summary of the rational choice perspective can be found in Cornish and Clarke (1986). Offenders are viewed as acting as a result of prior thought and judgement and as a result of weighing costs and rewards. It is acknowledged that their decision-making will be limited by constraints of time and the availability of relevant information. It is also acknowledged that the way in which factors are weighted almost certainly involves taking short-cuts in order to simplify the complex process of balancing advantages and disadvantages. Nevertheless, the approach presumes a level of rationality and decision-making in offending which in modern criminological discourse has been underplayed or ignored.

The research into burglars' perceptions and decision-making drew heavily on this developing perspective and its presumptions of rationality, decision-making, and choice. Within this perspective, opportunities for crime were regarded not simply as open windows or unlocked doors, but as comprising a balance of a variety of favourable and unfavourable situational factors, including the risks involved, the likely rewards and the ease or effort of committing the offence. The approach also recognized the role of opportunity costs in the decision to offend. In theory, at least, the attractiveness of committing a particular offence against a particular target might be weighed against the attractiveness of alternative targets and possibly even against the attractiveness of non-criminal pursuits.

The aim of the research, therefore, was to discover the way in which offenders made decisions about offending. In particular, it sought to determine which situational cues burglars used in their assessment of buildings as targets and the way in which these were weighed to arrive at a decision to offend. Apart from filling a gap in the research on the behaviour and perceptions of offenders, it was hoped that the result might also be useful in the development of rational crime-prevention strategies.

Research methods and samples

The main method, which involved the largest sample of offenders and which occupied the major proportion of the research time, was a semi-structured interview. The interviews provided a great deal of information on the way in which offenders assessed situational factors in their decision to offend. There were a number of important disadvantages, however, with using interviewing methods in this context. In particular, it was nearly always necessary for the interviewer to identify specific cues before they could be discussed with offenders. This carried with it the danger that the interviewer was generating concepts that had little relevance to the interviewees (even though they might oblige the interviewer by using them in their responses). In order to overcome this, a second method was devised which allowed the possibility of assessing offenders' perceptions of environments of crime without recourse to verbal prompting. This method comprised a video-recording of houses which was shown to burglars who were invited to comment on what they saw.

Different samples of offenders were used for each method. The semi-structured interview was administered to 128 offenders currently serving sentences in prison or borstal for burglary, plus a small number currently on probation. The video-recording of houses was shown to 40 offenders currently serving sentences in prison or borstal for burglary. Approximately half of each sample was aged under 21, all were males and over three-quarters were serving prison sentences. Most offenders were experienced in terms of both offending and frequency of their involvement in the criminal justice system.

The video-tape method

The video-tape method was designed to discover which situational cues burglars used in their assessment of potential targets. The method had an advantage that offenders could be prompted to articulate their perceptions of environments of potential crimes without any verbal prompting from the researcher. Conceivably, offenders could have identified any aspect of the situation confronting them. The cues actually chosen, therefore, were of particular significance.

A video-recording was made of thirty-six dwellings,

comprising eight blocks of three or four neighbouring houses and one block of three flats. The recording was made by filming from a van travelling along the road at approximately walking pace. Our intention was to present as much of each of the potential targets as might be revealed to an offender walking along the same route. Shots up and down the road were included at the beginning of each sequence of houses to supply information about the area. The offenders were told that they were to assess each of the dwellings as a potential target and to arrive at a decision about whether or not they would choose it. They were encouraged to think aloud and to tell us which factors they were considering in their assessment. Their comments and decisions in relation to each house were tape recorded and transcribed.

On average, 40 per cent of all offenders viewing a particular house decided that it was suitable for burglary. There was considerable variation, however, in the percentage of offenders who found particular houses acceptable. Only 8 per cent of burglars thought that a poorly kept, semi-detached council house was suitable, compared with 84 per cent who thought a large, well covered, detached house was suitable. A similar variation was found between the different blocks of houses. On average, semi-detached, council houses were found desirable by only 20 per cent of offenders, compared with detached, well-covered houses which were found acceptable by 70 per cent.

There was no significant difference between offenders who found a large proportion of houses suitable and those who found a small proportion suitable in terms of age, although there was a significant difference between them in terms of the total number of admitted burglaries. Those who admitted the greatest number of burglaries found the greatest number of dwellings suitable. This might be the result of experience generating the confidence to move away from familiar targets to unfamiliar ones or it might be the result of offenders who are willing to select almost any kind of target committing more crimes than those who are more limited in their choices.

Offenders' comments made while viewing the video-recording were tape recorded and analysed by selecting discrete statements and listing them. It is interesting that many responses did not relate directly to what was seen on the screen, but involved imputations about the type of dwelling or the type of people who lived there. For example:

> 'Judging by the fact that there's a truck parked in the garden, they've probably got their own business. It seems to be the sort of house for a businessman, someone who's out at meetings all day.'

> 'That house is so big. There's no way in the day that they can occupy all of that house. More than likely the husband would be working and the kids would be at school and a married wife out shopping.' (3)

The analysis of the lists of discrete statements revealed that over 90 per cent of concepts related to the three major problems that burglars have to face when deciding to commit a burglary: whether they can get away with it (risk), whether they can make anything out of it (reward), and whether they can do it (ease). Overall, risk factors were used by a greater percentage of offenders in their assessment than cues relating to reward or ease. When houses were found suitable for burglary, approximately half of the offenders, on average, noted at least one favourable risk factor in their assessment, about one-quarter of the offenders mentioned that rewards were favourable and about the same percentage said that there were favourable cues relating to the ease of entry. When houses were found to be unsuitable, almost two-thirds of the offenders mentioned unfavourable risk factors, one-fifth reported unfavourable reward factors and fewer than 10 per cent noted difficulties relating to entry.

It is worth restating that the offenders could have spoken about any feature of the immediate environment of the dwelling that could be seen on the video-recording. Conceivably, they could have noted the number of windows, the height of the building, the type of roof tiles or the colour of the doors. In fact, they spoke of none of these things. Most burglars focused their attention on the same small number of factors. The remainder of this section comprises an outline of these most frequently reported cues identified as important in their decision-making.

Concepts relating to risk

The most frequently noted risk factors concerned the amount of cover offered by the dwelling and its immediate environment. Comments relating to cover were wide-ranging. Apart from the expected references to trees, hedges, fences, bushes, and walls, offenders also mentioned slightly protruding garages, porches and even posts or saplings in the garden. Sometimes these objects would be narrower than the subjects themselves. It was argued that almost any obstruction, no matter how narrow, might be sufficient to break up their outlines and make them less visible:

> 'This porch, see that little porch there. You've got the porch at your back and you've got a little bit of cover. They have glass panels in the front door there. If the wall is coming out slightly, you've got somewhere to stand in while you are looking around.'

Offenders also frequently noted the closeness of neighbouring houses to the building being assessed. Terraced houses were often thought undesirable because of the proximity of one house to another. One mid-terraced house, for example, was described by almost half of the offenders who found it unsuitable as unfavourable on the grounds of the closeness of neighbours:

> 'I wouldn't do that, it's too close, they'd know one another and notice strangers.'

> 'Too near other property. It's the sort of property where you've got people nearby. They've probably got nosey neighbours.'

Implicit in these statements is the fear that neighbours will know a stranger to the area and will be more likely than passers-by to do something about it. For this reason, semi-detached houses were often disliked as much as terraced houses. Semi-detached and terraced houses were also disliked because the neighbours might hear the intruder through the adjoining wall:

> 'I don't think I would do that one because it's a semi. If they hear you moving around and they know there's nobody next door. They can hear people in the rooms in these type of houses, so they might come round to see who it was.'

It should be noted that the stated dislike of burglars for terraced and semi-detached houses was not reflected in the findings of the two British Crime Surveys (Hough and Mayhew 1983; 1985). It would appear that this was a result of other factors and other considerations influencing offender decision-making not covered by the video-recording. It is possible that the video-tape method reveals not actual but ideal choices of burglars. Offenders might take into account other considerations such as the distance to travel or their willingness to make an effort to travel on any particular occasion. In reality, ideal houses might be a long way from the offender's area of residence and a target less than ideal might have to be accepted.

Offenders often mentioned the relationship of the dwelling to the road. Such comments often concerned the distance of the house from the road. Being set back was frequently mentioned as a desirable feature and houses close to the road, such as terraced houses, were thought undesirable.

> 'It's too close to the road. Anybody walking past could watch over you. No, it's on the main road and it's a busy road.'

The main problem noted concerning the closeness of the road was the nearness of passers-by and drivers in passing vehicles who might see the offender. In addition, busy roads caused problems in relation to parking the getaway vehicle.

Another major concern among the offenders was whether the dwelling was overlooked by houses immediately opposite or directly behind. Again, the fear related to the prospect of being seen by neighbours who were familiar with the property and who might intervene:

> 'No, I wouldn't do that one because there are plenty of people in those houses who can see into the back garden. They just look out of their bedroom window or something and see you standing there trying to smash a window. They'd phone the police and you'd be caught straight away.'

There were no occupants actually visible on the video-tape, so all comments about occupancy related to signs of occupancy rather than to the sighting of people. These comments almost invariably were based on seeing a light on in the house or a vehicle parked close by. Some offenders were unable to articulate the exact cue that triggered the feeling that someone was in. By far the most frequently mentioned sign of occupancy was the presence of a car in the drive. The sight of a car near to a property would often be sufficient in itself to make the offender decide that the house was unsuitable for burglary. Many offenders suggested that they would have chosen a particular house if the car had not been there. As it stood, the car put them off:

> 'There's a car in the drive, so I wouldn't attempt it. I'd do that if the car wasn't there. Yeh, I'd have done that. It's big, it's off the road and bushes around.'

The only other sign of occupancy mentioned was a light on in the house. As the video-tape was filmed during the day, only one of the houses had a light on. Nevertheless, this was offered as a reason for not choosing that house by half of the offenders who found it unsuitable.

Offenders also commented on the existence or non-existence of escape routes. This usually meant additional exits at the rear or sides of the property rather than in relation to the area as a whole. Rear or side exits were considered desirable by some offenders because they offered an alternative means of getting away if anything went wrong:

> 'Yeh, I'd take that one. It's set at the back of the trees and there are plenty of ways to get out.'

> 'There's plenty of ways out if you had to get out quick, you know.'

The absence of escape routes was mentioned as a problem associated with flats. Almost one-third of the offenders who found a block of high-rise flats unfavourable mentioned their inability to escape quickly as a reason for not selecting them:

> 'I don't like flats. You could be on the tenth floor. If someone comes in, what are you going to do: climb out of the window ten floors up?'

> 'I wouldn't do that 'cos once you're in the flat, you are there to stay. There's no way out, you're caught straight

away.'

Many offenders mentioned the advantage of being able to get to the rear of the property, not as a means of escape, but as a means of gaining access to the back of the dwelling. It is possible that some offenders prefer the rear of the property because rear doors are more vulnerable or because rear windows are more likely to be insecure. However, most of the burglars explained their preference in terms of the cover provided by the dwelling itself and the reduction in risk of being seen by people at the front. Terraced houses were often criticized on the grounds of the difficulty of obtaining access to the rear.

> 'It's a terraced house. You can't get round the back. People would see you at the front.'

Detached houses offered a greater number of potential access points:

> 'It's a nice big house. There's numerous ways round the back. Definitely have a go at that.'

Only one of the thirty-six had an alarm visible, which meant that overall very few offenders mentioned alarms. Nevertheless, all of the offenders who found this house unsuitable for burglary pointed out the alarms as an unfavourable factor. Other cues were noted even less frequently. Offenders sometimes suspected that the occupant owned a dog, although no dogs were visible on the video-tape. Some made an attempt to guess the distance of the house from the nearest police station and made assumptions about the time it would take the police to arrive if called.

Concepts relating to reward

Comments relating to reward could not easily be subdivided. In most cases the offender made assumptions, on the basis of cues relating to the house, garden, and immediate area, about the wealth of the occupants and the likelihood of cash or goods being in the dwelling:

> 'That's all right. It looks prosperous, might be a five-bedroomed house. They might have money. I'd consider

that one.'

Although perceived affluence was noted as being an important factor in determining likely reward, the condition of the property also provided offenders with some insight into what might be in the house. It was assumed that if the occupants took a pride in the outside of their property, they probably took a similar pride in the interior and adorned their home with attractive and sellable goods. Houses that looked as if little attention had been paid to them were generally disliked:

> 'I wouldn't do that 'cos the car's in the garage and the curtains are closed. It looks a bit scruffy. I think they've let the home go a bit. There's nothing in that house but cockroaches. It just isn't the sort to do. You are not likely to find anything worth having to make it worth your while for the risk you are taking.'

Another factor which indicated likely rewards was whether the house was owned by the council. When this appeared to be the case the objective value of the house became relatively unimportant. For many offenders, council houses represented poverty:

> 'That looks like a council home. They haven't got any money.'

Offenders' clear dislike for council properties represents another departure from the findings of the British Crime Survey which showed that council homes were at greater risk of being burgled than non-council. One reason for this might, once again, be the disparity between the ideal world and the real world as mentioned earlier. Another reason might be the fact that the samples comprised older, more experienced offenders. Younger, less experienced offenders might be more likely to offend against council properties, which could have the effect of increasing the overall victimization rates for this kind of dwelling.

Concepts relating to ease of entry

As with the previous category, concepts relating to ease of entry could not easily be sub-divided. Statements were mainly non-specific and related to the general ease or difficulty of entering a particular building. Some comments, however, were directed at the nature or condition of the windows or doors. The size of the windows was also mentioned as affecting the ease of entry. Small windows were usually preferred to larger ones because they were easier to break or open. Sash windows were an exception as they often could be opened by slipping the old-fashioned catches with which many are fitted. Judgements about doors usually concerned the strength or the quality of the locks:

> 'This one, the door is quite easy to take off its hinges you know. By the looks of it, as far as I can see, it's only got an inch and a half lock in it. So the screws are about this long. I can't really see it clear enough. If that's just one lock, all you have to do is put a hand in the door and spring it.'

The semi-structured interview

The semi-structured interview provided the opportunity to discuss situational factors with offenders in greater detail. During the interview, the burglars were questioned about their perceptions of a number of specific situational factors and the influence that these had on their choice of targets. Each of the burglars was asked a question in the form, 'When you were last offending, would you have been put off a dwelling that had/did not have (a specific situational factor)?' Seven factors were discussed: occupancy, alarms, neighbours, whether overlooked, presence of dogs, passers-by, and locks.

The situational factor mentioned by the greatest percentage of offenders as capable of deterring them was occupancy. Over 90 per cent of them said that they would have always or sometimes been 'put off' by an occupied building. The presence of dogs was mentioned by three-quarters of offenders as capable of deterring them and just under three-quarters of the offenders said an alarm would have deterred them. Two-thirds said that they would have been deterred by the presence of neighbours and just over one-half by the property being overlooked at the side or rear. In contrast, only one-third of offenders said that they would have been put off by additional security locks and just over one-third said that they would have been put off by the presence of passers-by.

It was rarely possible for an offender to determine conclusively whether a house or a building was actually occupied from cues visible from the street. Offenders have to rely, in the first instance, on signs of occupancy rather than the sight of occupants. After the primary decision to approach the building had been made, offenders usually made some attempt to confirm their suspicion that the building was vacant. In the case of burglary in a dwelling, occupancy was usually determined by knocking on the front door, looking through a window or, less frequently, by telephoning the occupant.

Few offenders offered specific reasons for avoiding premises that were occupied. It seemed that to them the reasons were obvious: occupancy increased the chance of being seen, heard, or confronted by someone who had a vested interest in doing something about it. Some offenders said that they avoided occupied dwellings because they feared confrontation on the grounds that they might be forced to hit the occupant which to some of them was not only abhorrent, but also was seen as increasing the likelihood of getting caught and the seriousness of the offence. Some also mentioned that they feared frightening an old person who might die as a result, which once again was not only abhorrent, but also increased the seriousness of the offence. Very few offenders said that they would commit a burglary in a building that was clearly occupied. Most of those who said they would, specialized in night-time burglaries. Nevertheless, they often qualified their comments with the condition that they would only burgle a dwelling when the occupants were in bed asleep. Some offenders pointed out that, in these circumstances, there were very few risks and, as a result, there was little difference in offending at night in an occupied dwelling than in the day in an unoccupied one.

Most burglars said that they were deterred by burglar alarms. Those who were always deterred usually said that they simply knew nothing about them and had insufficient skill to overcome them. Others said that they would sometimes burgle a house with an alarm if the alarm could be by-passed or if it was simple enough for them to deactivate. The type of alarm was often an important factor in the decision. Their usual concern was with complicated or sophisticated alarms and with those which covered a number of potential entry points. About one-quarter of offenders said that they would burgle a property with an alarm. The most frequent reasons given for doing this were: they could 'leg it out of there' if the alarm sounded; or 'nobody would do anything about it anyway'. Others believed that many alarms are not properly activated and offered little deterrent value as they probably would not go off. Only three burglars said that a burglar alarm tempted them in that it suggested there was something in the house worth stealing.

Most offenders were deterred by the presence of neighbours. Some of them said that they only chose houses that were distant from other dwellings because of the threat of neighbours. Some said that they would only be deterred by neighbours if they had been seen by one of them. Most, however, said that they were deterred by signs that the neighbours were at home, using roughly the same cues as were used to determine whether the occupants of the target dwelling were at home. Fear of neighbours stemmed from the belief that they were nosey and that neighbours knew the activities of the occupant and knew what kinds of people regularly visited them. A central feature of this fear was the belief that neighbours, unlike other participants, would take some action if they spotted anything suspicious.

Whether or not a potential target was overlooked was also an important consideration. About half of the offenders who said that they would not commit a burglary against a target that was overlooked were concerned about the proximity of other buildings and whether there were windows facing potential entry points. The other half said that they would only be deterred if there was evidence of people in the overlooking properties or if there was someone actually watching them. The fear of people in overlooking properties was described in roughly the same way as was the fear of neighbours. People living or working in the area had a better knowledge of who was a stranger and who was supposed to be in the proximity of the target dwelling. People living nearby were also more likely to take action than passers-by. Offenders who were not concerned about overlooking properties held the opposite view that people in overlooking buildings would not do anything even if they did see someone acting suspiciously. Some of them admitted, however, that they would be more careful committing an offence against a dwelling that was overlooked.

The vast majority of burglars said that they were put off by dogs. Three reasons were offered for avoiding dogs: the most frequent reason was that dogs might bark and alert either the occupants or the neighbours; the second reason was the fear of being attacked by the dog; and the third was being placed in a situation in which it might be necessary to harm the animal. A number of offenders said that whether or not they were deterred depended on the type of dog. The breeds of dog most feared by this group were Dobermans and Alsatians. Others spoke only of 'large dogs' or 'yappy dogs'. Offenders who were not concerned about dogs felt that they could easily quieten the dog by feeding it, by talking to it or, if necessary, by killing it. Some felt that dogs were not generally very good at guarding properties and one described an instance in which a dog fell asleep in the same room in which he was searching.

Only a minority of offenders said that they were deterred by passers-by. Some of them said that they always avoided dwellings with people nearby and others said that whether or not they were deterred depended on circumstances. One condition mentioned was the number of people about. Some offenders said that when the number of people reached a certain level, they felt that there were 'too many people about' and looked elsewhere or called the offence off. Another condition mentioned was whether or not the people were stationary. It was argued that people walking past were generally thinking about other things and were not as observant as people standing still. A third condition was whether or not the passers-by had seen them or were looking at them. It was particularly worrying if the person held the offender's gaze. Those not deterred by passers-by typically said that they were only a threat when the property was entered from the front, which could usually be avoided. It was argued that once at the rear of the property, they could not be seen even if the street were crowded with people. Another argument was that passers-by would not do anything even if they did see you.

Additional security locks were seldom mentioned as deterrents. Those that said they were always put off by security devices generally argued that there was little point attempting to enter a secure dwelling when there were many more around that were less secure. Some offenders said that locks did not always put them off, but they were put off by them sometimes. An important factor was the type of lock and how difficult it was to overcome it. Chubb locks were frequently noted as being difficult locks to overcome and, hence, locks to avoid. Those who mentioned this, felt that it was pointless spending a long time attempting to enter a house when an easier one was almost certainly nearby. Others noted that spending too long on attempting to overcome a lock increased the chances of getting caught. The vast majority of offenders, however, said that locks did not deter them and stated that the type of lock was not a factor in their decision-making. Two main reasons were given for this: the first was that any simple levering tool could be used to break off most locks, no matter how sophisticated the manufacturers claimed them to be; the second was that they could simply by-pass them and find an alternative means of entry. A fairly simple alternative method of entry was to break a window.

Summary and conclusion

The research showed that the decision to offend or not to offend was most influenced by cues relating to the risk of getting caught. Cues relating to the potential rewards of the offence or the ease or difficulty of entry were mentioned less often. The most important risk factors concerned surveillability and signs of occupancy. Dogs and alarms were mentioned frequently by offenders and appeared to have the effect of occupancy proxies, standing in for occupants in their absence. Neighbours also represented an occupancy proxy in that they were seen as capable of acting on behalf of the occupants in their absence.

A great deal of the accumulated research evidence supports the conclusion that the key situational cues used by burglars relate to surveillability and signs of occupancy. Studies by Waller and Okihiro (1978), Winchester and Jackson (1982) and Reppetto (1974) and the British Crime Surveys (Hough and Mayhew 1983; 1985) all report differences in occupancy patterns among householders of victimized and non-victimized dwellings. Other research based on interviews with burglars also confirm the importance of occupancy in offender decision-making (Walsh 1980; Maguire and Bennett 1982).

It is surprising that cues relating to surveillability and occupancy have not featured more prominently in burglary-prevention programmes. One of the most frequently employed programmes is still the security campaign (e.g. the recent Magpie campaign). Crime prevention officers almost invariably advise householders to fit additional security locks, but they less frequently advise residents about surveillability or signs of occupancy. There is, however, some indication of a trend towards more diversified crime-prevention advice and suggestions relating to surveillance, the advantages of dogs and the involvement of neighbours in the creation of signs of occupancy are more common now than they once were. Rational burglary-prevention programmes need to be based on what is known about offenders' decision-making.

The relationship between offenders and the environment of crime is a complex one and by its nature cannot be elucidated by the efforts of just one study. Like any single piece of research the current research has managed only to scratch the surface. A great deal more needs to be known about the way in which offenders perceive potential targets and the immediate environment of crime and the way in which these perceptions shape their choices. It is particularly important, therefore, to establish a body of research which can be assessed as a whole. On the basis of studies which have already been conducted, some of which are summarized in other chapters in this book, it seems clear that the most profitable course is for this body of research to be multi-disciplinary.

Notes

1. For a theoretical overview of the situational approach, see: R.V.G. Clarke, 'Situational crime prevention: theory and practice', British Journal of Criminology 20, (1980), 136-47.

2. For a summary of the work of the HORPU on opportunities in crime, see: R.V.G. Clarke and P. Mayhew, Designing out Crime, Home Office Research Unit, HMSO, London, (1980); and K. Heal and G. Laycock, Situational Crime Prevention: from Theory into Practice, Home Office Research and Planning Unit, London, (1986).

3. Quotations presented in this chapter are selected from T.H. Bennett and R. Wright, Burglars on Burglary: Prevention and the Offender, Gower, Aldershot, (1984).

References

Bennett, T.H. and Wright, R. (1984) Burglars on Burglary: Prevention and the Offender, Aldershot: Gower

Clarke, R.V.G. (1980) 'Situational crime prevention: theory and practice', British Journal of Criminology, 20, 136-47

Clarke, R.V.G. and Mayhew, P. (1980) Designing out Crime, Home Office Research Unit, London: HMSO

Cornish, D.B. and Clarke, R.V.G. (1986) The Reasoning Criminal: Rational Choice Perspectives on Offending, New York: Springer Verlag

Heal, K. and Laycock, G. (1986) Situational Crime Prevention: from Theory into Practice, London: Home Office Research and Planning Unit

Hough, M. and Mayhew, P. (1983) The British Crime Survey, Home Office Research Study No. 76, London: HMSO

Hough, M. and Mayhew, P. (1985) Taking Account of Crime: Key Findings from the 1984 British Crime Survey, Home Office Research Study No. 85, London: HMSO

Maguire, M. and Bennett, T.H. (1982) Burglary in a Dwelling, London: Heinemann

Reppetto, T. (1974) Residential Crime, Cambridge, Mass.: Ballinger

Waller, I. and Okihiro, T. (1978) Burglary: the Victim and the Public, Toronto: University of Toronto Press

Walsh, D. (1980) Break-ins: Burglary from Private Houses, London: Constable

Winchester, S. and Jackson, H. (1982) Residential Burglary, Home Office Research Study No. 74, London: HMSO

Chapter Ten

SOCIAL RELATIONS, NEIGHBOURHOOD STRUCTURE, AND THE FEAR OF CRIME IN BRITAIN

Susan J. Smith

Fear of crime is of increasing concern to a range of academics and policy-makers who have more traditionally been preoccupied with the extent of deviance itself. A variety of evidence about the incidence and character of fear in the developed world (especially in the USA and, to a lesser extent, Britain) has been subject to scrutiny over the last decade, and the major findings have been widely reviewed (e.g. Baumer 1978; Clemente and Kleiman 1977; Garofalo 1979, 1981a; Garofalo and Laub 1978; Skogan and Maxfield 1981; Smith 1987). It is generally agreed that fear connotes more than simply awareness about crime, and more than concern about the problem of local deviance (though these emotions are not necessarily unrelated). Rather, fear is a state of constant or intermittent anxiety: its effects reach beyond the prudent management of risk to impinge on public morale, individual well-being and the quality of social life.

A geography of fear?

Variability in the extent and distribution of fear has been observed at all spatial scales. As Lohman (1983) observes, its precise character at any one place reflects the cultural climate of a country, the social climate of a community and the emotional state of individuals. Fear of crime has been the subject of few cross-national comparisons, though differences between the USA and Canada, and between the USA and Britain, respectively, are mentioned by Brantingham *et al.* (1986) and Smith (1987) (see also Kerner

1978). This chapter is concerned largely with experience in England and Wales, and in particular with some findings of the first two sweeps of the British Crime Survey (BCS), (1) which, together, contain a full set of questions monitoring the incidence and effects of fear of crime (see Box et al. forthcoming; Maxfield 1984a, 1987).

Within England and Wales, the broadest meaningful unit at which fear might be examined is that of the region (the nine economic and planning regions consolidated by the local government reforms of 1974), as shown in Table 10.1. (2) The extent to which awareness about local crime, worry about victimization and feelings of insecurity impinge on the consciousness of the public is striking. In 1982 one-third of the population felt unsafe walking alone in their own neighbourhood, and three-fifths worried about the possibility that they or others in their household might be a victim of crime. Although the authors of the British Crime Survey have argued that fear of crime is not a national problem in Britain (see Hough and Mayhew 1983, 1985; Maxfield 1984a), the evidence of Table 10.1 indicates nevertheless that an extraordinarily large proportion of the population is affected by this one issue.

Perhaps even more notable is the extent to which crime encroaches on the perceptions of Londoners (those living within what were the bounds of the Greater London Council (GLC)). These residents are over-represented amongst those believing that crime is very or fairly common in their home neighbourhood and has recently been increasing. Such perceptions are coupled with the most extensive experiences of fear and worry in the country.

Additionally, it seems that the North, the North West and the West Midlands contain a disproportionate number of fearful residents. However, given accumulating evidence that fear is predominantly an urban phenomenon (Baumer 1985; Kennedy and Krahn 1983; Maxfield 1984a), it seems likely that these regional differences in fear partly reflect the pattern of urbanization. This association might be expected not only because the risks of victimization are highest in cities, but also, as will become apparent, because of what urban life implies for the structure and organization of neighbourhoods.

Notwithstanding intriguing differences in the character and incidence of fear nationally and regionally, it is increasingly recognized that fear, like crime itself, might most meaningfully be examined at the level of

Table 10.1 Fears and perceptions of crime in England and Wales

Region	Proportion of the population in each region who:			
	Feel unsafe[a]	Worry about victimization[b]	Believe burglary has increased[c]	Believe mugging and robbery has increased[c]
Wales	28	50	31	10
The North	39	66	43	25
North West Yorkshire	37	64	48	29
Humberside	33	59	44	19
West Midlands	39	65	45	21
East Midlands	27	58	28	10
East Anglia	26	50	26	15
South West	27	54	35	16
South East (Outer)	33	57	36	15
South East (GLC)	42	67	53	44
England and Wales	34	60	41	22
Weighted n	3,822	6,719	4,546	2,411

Notes

(a) People who feel a bit or very unsafe 'walking alone in this area after dark'.

(b) People for whom worry about 'the possibility that you or anyone else who lives with you might be the victim of crime' ranges from being a big worry to an occasional doubt.

(c) People who believe these crimes are now more common in their own residential area than they were five years ago.

Source

British Crime Survey 1982.

neighbourhood or community. The importance of this scale of analysis for understanding the incidence of offender behaviour and victimization is discussed by Reiss (1986) and by a number of other essays (Reiss and Tonry, eds, (1986)); and the notion of community 'careers' in crime has been further amplified by Bottoms and Wiles (1987) and Hope (1986a). The neighbourhood context of fear has been examined in North America by McPherson (1975), Maxfield (1984b), Podolefsky (1979), Skogan and Maxfield (1981), Taub et al. (1981, 1984), and Taylor et al. (1984); and in Britain both local crime surveys (see, for example, Kinsey 1984; Sparks et al. 1977; Smith 1986; Young et al. 1986) and the BCS draw attention to the disproportionately high levels of fear found in the inner cities and on the country's poorer council estates (Hough and Mayhew 1985). This literature suggests that there exists a threshold beyond which fear of crime transcends individual emotion to constitute a social 'fact' - a property of groups and sites, best examined in spatial terms (Taylor et al. 1984). The predictors of such fears have been examined in some detail.

Explaining the 'neighbourhood effect'

To an extent, and as might be expected, intra-urban patterns of fear follow the incidence of crime (see Hope and Hough 1987; Shotland et al. 1979; Warr 1980, 1982). The second sweep of the BCS shows that people who live in high-risk neighbourhood types (those labelled in Acorn (3) as 'high-status non-family', 'multi-racial' and 'poorest council estates') are most likely to feel unsafe. However, there is some debate concerning the extent to which fears and risks coincide (an account of which is given in Smith 1987), and it is certainly the case that, both amongst social groups and within neighbourhood units, fear is more widespread than either experiences of recent victimization or perceptions of neighbourhood risks (cf. Maxfield 1987). Fear is not, moreover, necessarily most pervasive amongst the most vulnerable social groups, or within the most crime-prone locales (see Taub et al. 1984; Hindelang et al. 1978; Maxfield 1984a). There is fairly wide agreement, then, that local crime rates, though important, are neither necessary nor sufficient predictors of the extent of fear in a given population (Hough 1984, 1985).

As a consequence of the discovery that ostensibly low-risk groups (especially women and the elderly) and low-risk neighbourhoods can experience relatively widespread fears (Baumer 1985; Riger and Skogan 1978; Warr 1984; Riger 1985) much of the literature on the neighbourhood context of fear has explored its potential independence from known crime rates (Agnew 1985; Henig and Maxfield 1978; Skogan and Maxfield 1981; Maxfield 1984b). Inevitably, this strategy has been criticized by those (rightly) concerned to emphasize that public fears are often accurately grounded in local risks (Kinsey 1984; Young et al. 1986). Nevertheless, such research has successfully isolated a range of social and environmental cues that operate independently of objective risk to translate local events and experiences into a potential threat of crime. These cues are wide-ranging, but may be grouped as follows.

The mass media (newspapers, television, gossip, and rumour) are known to exaggerate both the extent of crime and its severity (cf. Garofalo 1981b; Mawby and Brown 1984). This might contribute to a general feeling of anxiety in some locales. I have, however, argued elsewhere that the reporting practices of the provincial press are more important for heightening public awareness about crime than for inciting their fears (Smith 1985).

Second, fear may be enhanced or inhibited by the actions of law enforcement officials. Block (1971), for instance, has argued that fear of crime is, in part, indexed by fear of the police. Though the wider applicability of this notion, certainly in Canada, has been questioned by Krahn and Kennedy (1985), it may well be that certain forms of police presence (which perhaps are interpreted as an index of local criminal activity) can independently trigger public fears (Lewis and Maxfield 1980).

Third, fear may be linked with the state of the built and lived environment. It seems that a variety of neighbourhood 'incivilities' including litter, graffiti, damaged property, loitering vagrants or youths, and so on, tend to be interpreted as evidence of criminality. These can alert people to the prospect of deviance and may so heighten their sense of fear (Lewis and Maxfield 1980; Wilson and Kelling 1982). Finally, anxieties related to a deterioration in community life, to a sense of social isolation, and to discontentment related to poor service provision appear, under certain circumstances, to be displaced on to the problem of crime (Merry 1981; Hunter and Baumer 1982; Skogan and Maxfield 1981; Smith 1983, 1986).

Recently, the links between these factors have been drawn together in an emerging literature exploring the relationships between crime and neighbourhood change (see especially, Skogan, 1986; Taub et al. 1981, 1984; Taylor et al. 1986). It has been argued that the gap between fear and actual risk is filled by the anxieties generated when communities have insufficient resources to resolve the uncertainties associated with urban decline. Fear of crime, according to this line of reasoning, is more than a response to perceived risk: primarily, it is 'a judgement that the government and the social structure will not be able to provide the collective good of safety' (Taylor et al. 1986, 176). In short, fear of crime may be conceptualized as an expression of the sense of powerlessness and uncertainty that accompanies much of urban life (see Lavrakas and Herz 1982; Lavrakas and Lewis 1980; Lewis 1980). Investigating this problem of perceived powerlessness in the face of unpredictable local change, Lewis and Salem (1981) link high levels of fear directly with residents' limited capacity to control the future of their lives and environment. They observed following a study in Chicago (pp. 414-5) that: 'Neighbourhoods with political power, for example, appeared more capable of addressing local problems than did those without it; and this capacity often appeared to contribute to diminishing fear.'

By revealing the extent to which fear of crime is encouraged by environmental and social cues, analysts have made an important case for linking the management of fear with the operation of a range of urban policies outside the spheres of crime control and law enforcement. They have also uncovered an association between fear and the political and social resources communities require to gain a sense of control over the uncertainties associated with urban lifestyles. However, research concerned with the role of environmental incivility and displaceable anxiety, and more recent interest in the anxieties generated by disinvestment, deindustrialization, real estate management, demolition, and construction within cities (Taub et al. 1984; Skogan 1986), is not yet well integrated with evidence concerning the differential distribution of fear amongst households and individuals.

There is now ample evidence that fear varies in its extent according to age and gender (Braungart et al. 1979; Sundeen and Matthieu 1976; Feinberg 1981; Giles-Sims 1984; Kennedy and Silverman 1985a; Warr 1984), 'race' (Smith and

Gray 1985) and 'class' (Miethe and Lee 1984). This literature suggests that neighbourhood-based explanations of fear which underplay the effects of social relations within and between households will neglect some important facets of crime-related anxiety. Such 'top-down' approaches do not, for instance, readily explain why fear is sometimes differentially experienced by individuals sharing the same social and environmental milieu, nor do they show why the differentiating criteria in such cases often varies in space (though Maxfield, 1984a, has shown that fear in low-crime neighbourhoods is much more socially discriminatory than is fear in high-crime locales). More recently, Maxfield (1987) has argued that socio-demographic differences in fear are largely 'explained' (in a statistical sense) by similarly aligned differences in experiences of crime and perceptions of risk. Nevertheless, the task of accounting theoretically for the character and consistency of the social alignment of fear (and therefore of its predictors in experiences and perceptions) remains. To this end, it may be necessary to complement top-down neighbourhood-based approaches with a strategy building upwards from a structure of social relations encapsulated in the status and organization of households - a structure systematically affording some residents a greater sense of safety than others (both by rendering them more or less vulnerable to crime, and by impinging on their actual and perceived abilities to manage and control personal space or territory).

This essay makes the argument that, to an extent, generalizations about the capacity of neighbourhoods or 'communities' to handle fear must be qualified with evidence concerning the differential capacity of individuals and households to do the same thing. Variations in public abilities to control unpredictability at a personal level (related to perceptions of place and the practice of forward planning) have, for instance, already proved important in understanding the disproportionate fears of the elderly (Normoyle and Lavrakas 1984; Patterson 1978, 1979). This chapter is concerned with two elements of a broader set of local social relations which, for the purposes of examining the fear of crime, can be regarded as mediating between individuals and neighbourhoods.

Household structure and housing tenure figure prominently in the literature of geography and sociology as factors significant for an understanding of the quality and organization of social life. Each is a medium through which at least part of the residential environment gains an imputed meaning (including meanings associated with fear, anxiety, or a sense of safety), though they are frequently peripheral to neighbourhood studies of fear (DuBow and Emmons 1981). Theoretically, however, both household type and housing tenure as defined for the present analysis can be interpreted as an index of the uneven distribution of power: the former is bound into the structure of gender relations; the latter into the social relations of consumption. If collective powerlessness is a key to explaining why neighbourhood uncertainties become neighbourhood fears, powerlessness at the level of the household may be a key to understanding how the local expression of fear is socially constructed.

The relevance of household type and tenure for an understanding of the differential distribution of fear also has some empirical underpinnings. A logistic regression analysis using the 1984 BCS with fear (F) (4) as the dependent variable and nine independent variables, including a crude indicator of risk (R), (5) three measures of incivility (L, H, S), (6) two measures of residential satisfaction (RS, C), (7) age (A), (8) household type (F), (9) and tenure (T), (10) revealed household type as the variable with the largest effect on the odds of feeling unsafe. The effect of tenure is smaller, but comparable with that of incivility, and with the less powerful categories of age, risk, and residential satisfaction. The linear form of the model is given in the notes. (11)

To begin to consider the relevance of household type and tenure for the distribution of fear, the remainder of the chapter falls into three sections. The first simply describes the incidence of fear amongst gender-differentiated household types and between tenure groups (to economize on space, the latter discussion relates only to owner occupiers and council renters; these constitute 89 per cent of the total sample). This is followed by an exploratory examination of the factors likely to have encouraged such a distribution. Finally, some implications of the analysis for the management of fear in urban Britain are considered.

Households, housing and the impact of crime

Table 10.2 expands the four categories of households used in the logistic regression model to six (though generalizations

Table 10.2 Fear of crime within household types

	Proportion in household grouping who feel unsafe[a]			
	Neighbourhood risks:			
	Low	Medium	High	Weighted n
Single females	56	68	72	717
One-parent mothers	51	60	65	139
Family females[b]	42	49	63	4,551
Single males	16	24	25	402
One-parent fathers[c]	0	9	24	16
Family males[b]	9	14	24	4,549
All respondents	27	35	45	10,374

Notes

a People who feel a bit or very unsafe 'walking alone in this area after dark'.

b Respondents living in households with at least one other adult.

c The numbers here are too small for estimates to be meaningful.

Source
British Crime Survey 1984.

about single-parent fathers are precluded by small numbers). Although fears are consistently most widespread amongst the residents of high-risk neighbourhood types, it is also clear that within each category of risk, a disproportionate share of public anxiety is born by women living alone and by one-parent mothers (see Markson and Hess 1980). Fear is also marked amongst 'family females', (12) especially those resident in high-risk zones. (13)

For the most part, this differential distribution of fear within and between household types is not accounted for by patterns of recent victimization. Despite a high incidence of fear, women living alone, who comprise 7 per cent of the sample, experienced only 3.4 per cent of all personal offences recorded in the survey (viz. common assault, theft from the person, wounding and robbery), and they live in households with a below average likelihood of experiencing vandalism or burglary. Family females also experienced less than half the personal offences that might be expected given their proportion in the population. Only single-parent mothers seem over-represented as victims, experiencing four times the expected proportion of personal offences and twice the average incidence of burglary (though small numbers mean such a generalization should be treated with extreme caution).

Fear amongst men (whether in single or multiple adult households), is much less pervasive: men are between one-third and one-half as likely as women to feel unsafe. Men's fears are most widespread amongst those living alone, though this discrepancy disappears amongst the residents of high-risk neighbourhood types where about one-quarter of all male respondents (as compared with between two-thirds and three-quarters of women) feel insecure.

Much the same pattern of gender-differentiated household differences that appears in relation to perceptions of safety is also apparent regarding personal worries about the risk of being mugged and robbed, though the gap between men and women is smaller (one-fifth of single men in low-risk neighbourhood types, rising to one-third in high-risk areas, are very or fairly worried about this crime type; the corresponding proportions for family males are one-quarter and two-fifths), and single-parent mothers living in low-risk areas evince particularly marked anxiety (54 per cent of these, as compared with 42 per cent of single women and 47 per cent of family females, are worried). The differences are much less marked for burglary, the threat of which prompts half the respondents in each neighbourhood type to worry (though again such anxiety remains most widespread amongst women, affecting 70 per cent of family females as compared with 62 per cent of family males in high-risk zones).

Finally, it is worth pointing out that most differences in worry and anxiety about crime persist irrespective of overall levels of residential satisfaction. Even amongst those who claim to be 'very satisfied' with living where they do, over half the women living alone and over one-third of women living in other household types feel somewhat unsafe walking alone locally after dark.

The consistency of housing tenure as a predictor of fear and anxiety about crime is also marked. Table 10.3 shows that for all categories of risk, around 10 per cent more

Table 10.3 Housing tenure, perceptions of safety, and anxiety about crime

	Proportion who feel:					
	(i) Unsafe[a]	(ii) Worried about:[b]				Weighted
		Burglary	Vandalism	Mugging	Attack	n
Low-risk[c]						
Own	26	54	48	35	32	5,081
Rent[d]	34	51	49	41	39	819
Medium-risk						
Own	29	65	58	44	40	1,303
Rent	41	63	62	54	50	1,298
High-risk						
Own	42	66	59	52	47	640
Rent	51	73	72	63	56	404
All						
Own	28	57	51	38	35	7,024
Rent	41	60	59	51	47	2,520

Notes

a Those who feel a bit or very unsafe 'walking alone in this area after dark'.

b Those who feel fairly or very worried about 'having your home broken into and something stolen', 'being mugged and robbed', 'having your home or property damaged by vandals' or 'being attacked by strangers'.

c Neighbourhood risks as defined in footnote 13.

d Council tenants only.

Source

British Crime Survey 1984.

tenants than owners express a sense of insecurity. Tenants are also more likely to feel fairly or very worried about being mugged and robbed, or attacked, insulted, or bothered by strangers. Although fears about property crime do not evince such a marked discrepancy (except in high-risk neighbourhood types), there is evidence even here that the worries of renters are most acute. For instance, though around 2 in 5 owners and renters worry about burglary, a

Table 10.4 Fear of crime and residential satisfaction amongst owners and council tenants

Level of satisfaction with living in local area. Proportion in each tenure group who feel unsafe (a) amongst those who are	Very satisfied	Satis-fied	Dis-satisfied	Very dis-satisfied	Weighted n
Owners	24	29	42	45	2,013
Renters	34	40	54	66	1,053

Notes
(a) Those who feel a bit or very unsafe 'walking alone in this area after dark'.

Source
British Crime Survey 1984.

larger proportion of renters (29 per cent as compared with 21 per cent of owners) claim to be '*very* worried'. Renters are also more likely than owners to believe they are certain or very likely to experience a range of personal and property crimes, though relatively few respondents hold such beliefs (the proportions range from 1.7 per cent - for owners' assessments of their probability of being attacked, to 12.9 per cent - the proportion of renters who expect to be burgled). Renters' perceptions in this respect may well be an accurate reflection of their actual vulnerability (see Hope 1986b; Hope and Hough 1987), but the fact that their fears remain disproportionately high even when they live in low- or medium-risk neighbourhood types (the greatest discrepancies in the fears of owners and renters are found in 'modern family housing, higher income' and 'poor-quality older terraced housing' enumeration districts, which are low- and medium-risk, respectively), and that they persist irrespective of the level of residential satisfaction (Table 10.4) suggests that other factors need to be taken into account.

The salience of gender and tenure: exploring some hypotheses

This section considers some factors other than neighbourhood crime rates and individual risk that might account for the differential distribution of fear within and between household types. My aim is to develop the thesis that patterns of fear are associated with the availability or otherwise of the resources individuals need to gain a sense of control over their life paths and territory. I suggest that such resources are a function not only of neighbourhood politics and economy (factors that clearly are important, but already relatively well-documented in the literature) but also with inequalities in the distribution of power that are associated with: (i) the structure of gender relations (indexed by household type); and (ii) the social relations of consumption (indexed by tenure). The analysis should be regarded as a preliminary exploration of a much broader topic. It provides an overview of a small part of a large survey; it has not yet been informed by the rich insight that qualitative case studies of fear of crime in Britain might yield. What follows, therefore, raises almost as many questions as it resolves about the structures of social relations in which anxiety about crime is embedded.

Households, families, and the fear of crime

Gender is obviously a key factor controlling the incidence of fear within and between households, and it has been built into the household type variable throughout the present discussion. Previous empirical research nevertheless dictates that the role of age must be a first consideration in explaining the salience of a gender-differentiated measure of household composition. There is an extensive literature, based especially on experience in the USA, documenting the importance of age as a predictor of both concern and anxiety about crime (see, for example, Clarke and Lewis 1982; Jaycox 1978; Miethe and Lee 1984; Kennedy and Silverman 1985a, 1985b). The disproportionate fears of the elderly in Britain have also been noted by Kinsey (1984), Maxfield (1984a) and Hough and Mayhew (1985). Since a relatively large proportion of elderly households consist of single, widowed, or divorced women, and given that amongst the elderly it is often single-adult householders who are most fearful, age seems likely to account for at least some of the variations in fear between different household types.

Table 10.5, however, offers only limited support for the age hypothesis. Certainly fear is most widespread amongst the elderly, but gender retains its salience as a predictor of fear amongst all age groups, and its discriminatory power

Table 10.5 Age, gender, and perceptions of safety

	Proportion who feel unsafe[a] amongst those ag				
	16-30	31-45	46-60	Over 60	Weighted n[b]
Females	42	42	48	60	5,617
Males	7	7	12	27	5,158
Single women[c]	48	55	48	66	890
Family Females[d]	42	41	48	56	4,727
Overall	25	25	31	45	10,776

Notes

a People who feel a bit or very unsafe 'walking alone in this area after dark'.

b Totals do not match those on Table 10.2 owing to different numbers of missing values.

c Includes single-parent and single-person households.

d Women living with one or more adults (with and without children).

Source

British Crime Survey 1984.

decreases rather than increases with age: whereas women are six times as likely as men to be fearful at ages 16-45, this over-representation declines to a factor of four between ages 46 and 50, and to a factor of two amongst the over-sixties (a similar interaction of age with gender has been observed in the USA, for instance, by Baumer, 1985). Moreover, amongst women in single adult households, fear is as widespread amongst the middle-aged as amongst the elderly. Age, then, is an important factor in considering the distribution of fear, reflecting not least the relative powerlessness of the elderly which limits the potency of their demands on crime-prevention policy (Berg and Johnson 1979) and denies them sufficient economic and political resources to provide for and control the future (Eve and Eve 1984). Age cannot, however, fully explain the salience of household-related variables in relation to the distribution of fear.

A second possibility, more directly linked with the structure of gender relations within households, is that fear of crime is enhanced by anxiety about the welfare of children. Certainly, the first BCS showed that many who worry about crime evince as much concern for others in their family as for themselves. Amongst those living in two- (or more than two-) adult households, therefore, the fact that fear is more widespread amongst women than men might reflect a division of family responsibilities in which women absorb the burden of fear related to their children's vulnerability. Consistent with this reasoning, a recent study by Herbert and Hyde (1984) has drawn attention to the high levels of fear expressed by women with children. The second BCS, however, reveals that living in households with dependent children makes no difference to the incidence of fear or worry about crime amongst male or female adults under sixty (though this does not rule out the possibility that the widespread fears amongst single-parent mothers are related to the strain of taking sole responsibility for the safety of their children).

Finally, the distribution of fear amongst household types may be interpreted as a reflection of gender differences in the 'gap' between perceived personal vulnerability and the options available to resist or avoid the threat of crime. Table 10.6 indicates that, with the possible exception of single-parent mothers, women's greater fears are not related simply to assessments of risk. Table 10.7 suggests further that although public fears are exacerbated by 'vicarious victimization' (i.e. knowledge of others who have been victimized locally) it is, for the most part, single men whose anxieties are most consistently inflated as a consequence of this indirect exposure to crime (though vicarious mugging and robbery also has a notable effect on the pervasiveness of fear amongst women in single-adult households). Similarly, both men and women take steps to avoid becoming a victim; and, surprisingly, a greater proportion of men (17.8 per cent) than women (10-15 per cent, depending on household type) avoid activities in which they would like to participate in order to minimize risk (though this generalization may not hold in the poorer parts of the major cities where surveys indicate that women's lifestyles have been severely affected by fear of crime (see, for instance, Kinsey 1984; Young et al. 1986).)

It is, nevertheless, amongst women generally, and women in single-adult households in particular, that awareness of and reactions to crime are most closely coupled with widespread fear. This disproportionate effect of perceived threat on women's fears is linked by Warr

Table 10.6 Perceived vulnerability to crime

Household type[b]	Proportion in each group who feel they are likely to be victims of[a]			
	Burglary	Vandalism	Mugging	Attack
Single females	33	26	19	17
One-parent mothers	50	39	28	24
Family females	42	32	19	16
Single males	32	24	13	12
One-parent fathers	29	29	13	17
Family males	34	23	12	11
All respondents	38	26	16	14
Weighted n	4,064	2,790	1,714	1,510

Notes
(a) Those who believe they are fairly likely, very likely or certain to become victims in the next year.
(b) These categories are equivalent to those given in Table 10.2. Note that small numbers make estimates for one-parent fathers unreliable.

Source
British Crime Survey 1984.

(1984) to the gender-differentiated impact of 'danger' signals in the environment. He argues that the heightened effect of these signals can, especially amongst younger women, reflect fear of rape or sexual assault, both of which tend to act as 'master offences', enhancing women's fears of many other kinds of crime (Warr 1985). Certainly, the BCS indicates fear of rape to be particularly pervasive amongst women aged 16-30, over 40 per cent of whom are 'very worried' about becoming a rape victim (Hough and Mayhew 1985). Amongst older women, however, Maxfield (1987) has suggested that fear of crime is related less to worry about specific victimization than to an ill-defined sense of unease (whose best predictor is environmental incivility, especially the presence of teenagers hanging around the streets). For both these groups, it can be hypothesized that the effect on fear of environmental danger signals (whether these are

Table 10.7 'Vicarious victimization' and worry about crime

	Proportion who are worried[a] amongst[b]				
	Single women	Family females	Single men	Family males	Weighted n
Vicarious[c] victim of:					
Burglary					
Yes	64	67	63	64	1,462
No	53	58	41	50	4,051
Mugging					
Yes	67	61	52	38	468
No	52	50	25	28	3,884

Notes

a Those who are fairly or very worried about 'having your home broken into and something stolen' or 'being mugged and robbed'.

b Categories as in Table 10.5.

c In this table, a vicarious victim refers to a respondent who personally knows someone who has been a victim of the given offence in the last year, and believes that the offence took place locally.

Source

British Crime Survey 1984.

rooted in specific threat or general unease) is related to the resources available to women to retaliate against or avoid the people, places, and events that threaten them.

Worrall's and Pease's (1986) analysis of the first BCS shows that although female victims of personal crimes are more likely than men to know the offender, their plight is only half as likely to come to the attention of the police. In view of the discrepancies between men's and women's anxieties, the prospect that women's fears are exacerbated by their actual or perceived inability to mobilize official responses to personal crime cannot be discounted. Table 10.8 allows a similar line of reasoning concerning the orientation of *informal* reactions to crime, and begins to account for differences in the pervasiveness of fear amongst women from different household types. For instance, in

Table 10.8 Reactions to crime

Precaution[b]	Household type (% who make response)[a]				
	Single women	Family females	Single men	Family males	Weighted n
Avoid people	32	32	19	14	2,522
Avoid places	35	37	18	15	2,810
Take guardian	36	49	7	9	3,021
Avoid public transport	13	10	2	2	271
Use car	22	33	7	10	2,225
Miss events[c]	15	10	14	18	1,509

Notes

a Categories as in Table 10.5.

b People who always or usually take the given precaution when out after dark (except c).

c Those who say there are events and activities they would like to go to but do not because of crime and violence.

Source

British Crime Survey 1984.

developing strategies to minimize risk, women in multiple-adult families are able to make greater resort to cars and guardians than those in single adult households. The latter are less likely to have the option of avoiding public transport, and are more likely to be forced by the perceived threat of victimization to miss activities they would like to attend. The effect may be to reinforce their social isolation and so intensify fear (see Riger et al. 1978, 1982). This process is already well-documented amongst home-bound populations and others whose activities are limited by the perceived threat of crime (Kennedy and Silverman 1985b; Silverman and Kennedy 1984); its significance is not questioned by the disproportionate levels of fear experienced amongst women in single-adult households in Britain.

To summarize, three possible explanations for the differential distribution of fear amongst gender-differentiated household types have been considered: the effects of age, and the physical, and economic powerlessness (variously experienced by men and women) that so often accompanies it in modern Britain; the influence of children, and the extent to which the division of labour within households impinges on the distribution of worry about crime; and the differential availability between men and women and, more particuarly, between women in different household circumstances, of the resources required to manage perceived threat. Empirical evidence from the BCS offers tentative support to the first and third of these explanatory clusters. The analysis does not reject the possibility that fear expresses a socially structured pattern of inequalities in the public's capacity to control the threat of crime.

Fear and the social relations of housing tenure

A second area of concern derives from the fact that crime impinges more on the lives of council tenants than owner occupiers. Currently, the majority of theoretical interest in examining such differences in Britain centres on recent changes in the structure and organization of the housing system. Though all post-war governments have favoured an extension of owner occupation, it is only during the 1980s that moves to reduce public spending in housing have been explicitly justified in ideological terms; for almost a decade now, British policy-makers have worked with the assumption that the market can house the majority of the population more adequately and efficiently than can the state. Thus between 1979-80 and 1985-6 public expenditure on housing declined from 5.8 per cent to 2.5 per cent of the social services budget (see Crook 1986) - a decline more marked than in any other social services, making changes in housing provision one of the most visible elements of the current restructuring of the welfare state in Britain. The corollary of this process is an extension of owner occupation (a trend greatly accelerated by the 'right to buy' clause of the 1980 Housing Act). Though the changes have involved a net shift rather than an overall reduction in state subsidy (savings on the provision of housing in kind are offset by increased payments on tax incentives associated with ownership) the role this assigns to council housing signals a marked break with the past (Robinson 1986).

In the wake of these trends in policy and politics, housing studies have become central to research on the social cleavages that derive from differential access to public and private modes of consumption (Forrest and Murie 1986; Saunders 1984, 1986). Increasingly, material, behavioural and attitudinal differences between owners and renters are being interpreted in the light of the 'residualization' of the public sector - a process whereby:

> public housing moves towards a position in which it provides only a 'safety net' for those who for reasons of poverty, age or infirmity cannot obtain suitable accommodation in the private sector. It almost certainly involves lowering the status and increasing the stigma attached to public housing.
>
> (Malpass and Murie 1982, 174)

'Residualization' is the product of an absolute and relative reduction in the size of local authorities' housing stocks, coupled with a selective pattern of sales that has increased the degree of socio-economic polarization between owners and renters. Bentham (1986) has monitored this process, showing that the income distribution of owners and renters steadily diverged between the mid-1970s and 1983 (to the extent that renters' median incomes fell from 75 per cent to 45 per cent of those of owners). Forrest and Murie (1983, 1986, 1987) document the trend towards the public sector housing an increasing proportion of those dependent on state benefits, especially the unemployed, and particularly the long-term unemployed (see also Forrest and Murie (1986, 1987)). These authors point out that, although the marginalized poor have always received the worst housing, the 1980s has seen an unprecedented level of concentration in the public sector of benefit and service-dependent populations - groups now increasing in size as a consequence of economic and employment changes. The salience of tenure as a predictor of the fear of crime must be viewed, therefore, not in terms of immutable attitudinal differences between owners and renters, but in terms of the changing social relations of consumption in British society which happened to be best expressed through housing tenure at the time the BCS was conducted.

With this in mind, it might be argued at one level that the disproportionate impact of crime on public sector tenants (through worry and anxiety, as well as through actual victimization) is a consequence of the concentration in that sector of high offender-rate populations (i.e. of those most likely to participate in the offences that are most vigorously policed and commonly sanctioned): the poor, the unemployed, children from 'problem' families, and so on. These facets of the population structure certainly seem to be reflected in high offender rates in parts of the public sector (Baldwin and Bottoms 1976; Bottoms and Wiles 1986); and (in so far as a high proportion of the crimes of such offenders are committed locally) in the higher victimization rate for some crimes amongst council tenants (Hope 1986b). The public, too, are aware of the greater overlap of offender and victim populations amongst tenants: the 1984 BCS indicates that 39 per cent of tenants but only 29 per cent of owners think the majority of local burglaries are committed by local residents.

The pervasiveness of deviance, and of its effects, in particular locales or parts of the social structure (in this case, amongst local authority tenants) traditionally invites a sub-cultural explanation of both attitudinal and behavioural differences between sub-populations. The notion that certain locales evince tolerance of and assent to the delinquent way of life proved helpful, for instance, to both Evans (1980) and Herbert (1982) in their attempts to explain the distribution of deviance in Cardiff and Swansea; and in many ways, a cultural explanation of the distinctions between owners and renters fits neatly with the residualization thesis outlined above. However, such an interpretation would not expect tenants' fears to be so much more widespread than owners', and it would anticipate a greater tolerance from the former than from the latter of a range of crime types. Yet Table 10.9 shows that, if anything, tenants view crime, particularly crimes committed for material gain, as more serious than do owners. (This suggestion that social class is not a key factor affecting views about crime seriousness is confirmed by Pease, forthcoming). Moreover, far from providing evidence of tenants' sympathy with crime-related lifestyles, it is clear from Table 10.10 that 'vicarious' experiences of crime, especially of mugging and robbery, exacerbate fear amongst tenants to a greater extent than amongst owner-occupiers.

An alternative 'cultural' perspective on the distinction between owners and renters might draw on the greater anonymity that can be associated with life in the public sector. Certainly, tenants are less free than the majority of owners to choose their neighbours, and Kail and Kleinman (1985), Mugford (1984) and Taylor et al. (1984) have all argued for the importance of friendship and support

Table 10.9 Perceptions of crime seriousness

Crime type[a]	Proportion who think crime very serious amongst: Owners	Council tenants
Someone being mugged and robbed	86	82
Fiddling income tax	19	24
Joy-riding[b]	29	33
Domestic burglary	59	61
Domestic vandalism	56	56
Fiddling Social Security	27	34
Attack by stranger	74	72
Weighted n	7,334	2,552

Notes

a A selection only from the range monitored by the BCS.

b A car being stolen for a joy ride.

Source

British Crime Survey 1984.

Table 10.10 Housing tenure and the impact of 'vicarious victimization'

	Proportion who are worried[a] amongst		
Vicarious[b] victim of:	Owner-occupiers	Council tenants	Weighted n
Burglary			
Yes	64	69	1,929
No	53	57	4,205
Mugging			
Yes	45	65	434
No	37	49	3,592

Note

a As Table 10.7

b As note c, Table 10.7

Source

British Crime Survey 1984.

networks in controlling personal fears about crime. However, the BCS elicits no differences, on average, between owners and renters either in the proportion of local people who are regarded as friends or acquaintances (about 30 per cent for both groups) or in the proportion (around two-thirds) who say it would be easy for neighbours to keep an eye on their home if they were out.

In view of the evidence, even from these rather crude indicators, it makes little sense to view council tenants as culturally distinctive in terms of their orientation towards delinquency. A more appealing starting point might be the reliance of tenants on a mode of state provisioning - public housing - that grants them limited access to the resources needed to manage the impact of crime. Consistent with the thesis linking fear with powerlessness, Forrest and Murie (1983) have aruged that the 'critical context for discussion of residualization involves issues of economic, political and social power' (p. 461). Certainly, it can be argued that those dependent on what Saunders (1986) calls state- rather than market- or self-provisioning are least able to control their lives and environment. The fact that residualization has been coupled with a decline in the quality of housing provided is also a symptom or consequence of the powerlessness of tenants to resist reductions in standards (see Forrest and Murie 1983; Malpass 1986). In the face of such trends, public support for owner occupation might itself be interpreted as a quest for 'ontological security', or the desire for control of, and safety within, one's personal or household space (cf. Saunders 1984).

Tenants are only slightly less likely than owners (31 per cent versus 34 per cent) to believe that target-hardening strategies increase household safety, yet for many council tenants, bureaucratic inertia coupled with low purchasing power diminishes their ability to acquire that sense of safety encouraged where homes are in good repair and properly secured. The inflexibility built into the allocation and distribution of public housing further limits tenants' options to move to a 'safer' neighbourhood when fears become intolerable. Amongst benefit-dependent tenants, moreover, economic hardship exacerbates the impact of material losses associated with property crime (a problem further compounded by the fact that only half the council tenants interviewed in contrast to four-fifths of the owner-occupiers have their homes insured). It is not difficult, therefore, to find explanations for the disproportionate fears of tenants that are rooted in their weak position in the social relations of consumption.

It is consistent with such explanations that amongst tenants there is a steady extension of feelings of safety as income increases (a perception not enjoyed by owner-occupiers whose sense of safety is most widespread in the lower middle-income bands). The possession of economic resources (an index of, amongst other things, potential control over crime-prevention strategies) seems, for tenants, to compensate for the fear-inducing effects of the social cleavage between them and the bulk of owner-occupiers. Unfortunately, while this kind of argument may be taken a little further by more detailed survey analysis, ultimately its verification or rejection will rest on a programme of qualitative research able to elicit the importance of tenure in relation to the meaning and symbolism of the home. Regrettably, the emerging literature on this theme has not yet addressed the issue of security from crime (see Duncan 1981; Holdsworth 1979; Porteus 1976).

The management of fear in urban Britain

There is some debate amongst academics and politicians concerning the extent to which fear of crime is a policy problem in its own right. In the USA, Brooks (1974) and Clemente and Kleiman (1977) imply that public anxieties do need active management, while Stafford and Galle (1984) argue that it is crime control that holds the key to alleviating fear. In Britain, where fear is generally regarded as a problem of local rather than national proportions, the grounds for special policies seem slim. Maxfield (1987) opts primarily for 'tell the truth' campaigns to provide the public with accurate assessments of risk. Mayhew (1985) also argues against more elaborate arrangements - less because they seem inappropriate than because they may be impossible, for 'if fear of crime is actually based on a wider concern about social change and disintegration, attempts to reduce fear may face even more intractable problems than attempts to reduce crime itself' (p. 90).

On the other hand, there is accumulating evidence that fear of crime may contribute to urban decline (Katzman 1980; Sampson and Wooldridge 1986), and there must be a point at which the impact of fear becomes more far-reaching than a society will tolerate. In this event, given the diverse origins of fear, the effectiveness of local crime-prevention strategies in increasing the quality of life could be enhanced by the introduction of selective measures to reduce environmental incivility and alleviate displaceable anxiety (strategies which could, at relatively low cost, be grafted on to existing forms of urban and social policy). It has also been suggested that the more debilitating effects of fear can be contained when local democracy is structured so as to provide the public with a sense of control over environmental improvements, social reforms, and so on (Lewis and Salem 1981; Smith 1987).

Notwithstanding the achievements of these general measures, this paper suggests that some important dimensions of fear are embedded in a structure of social relations extending beyond the boundaries of community or locality. Since such structures are unlikely to change at a rate convenient for those concerned with urban management, it seems reasonable to argue in the short term for allowing policing and urban policies to focus attention on the problem of fear amongst particular individuals or social groups. Currently, it seems especially pertinent to address: (i) the fears of women in the domestic sphere (where law enforcement officials are traditionally loath to interfere); and (ii) the disproportionate anxieties of council tenants (who often reside in areas where schemes like neighbourhood watch are least successful). This is not, however, an argument against continuing with neighbourhood- or community-based forms of crime prevention or urban management. It is, rather, a plea that the criterion of local sensitivity so often built into such schemes is orientated as much to the specific needs of individuals and households as to the aggregate characteristics of a physical and social milieu.

Notes

1. The first BCS sampled 11,000 people aged 16 or over in England and Wales (together with 6,000 in Scotland, who are not included in this paper), and has a response rate of 80 per cent. The 1984 survey interviewed a similar number of respondents, but excluded Scotland, and achieved a response rate of 77 per cent. The technical details of the surveys are given in NOP (1985) and Wood (1984). Unless

otherwise acknowledged, all references to BCS data in this chapter refer to the author's analyses. Small discrepancies with other BCS publications may be expected where different combinations of sub-populations give rise to different proportions of missing values.

2. An introductory outline of these regions' characteristics is given in House (1978). Their changing fortunes are monitored in Martin and Rawthorn (1986). Their crime rates are discussed in Smith (1986).

3. Acorn - A Classification of Residential Neighbourhoods (Produced by CACI) - is a system of classifying households on the basis of the census characteristics of the enumeration district in which they occur. The application of Acorn to the BCS is discussed in Hough and Mayhew (1985, Appendix F).

4. A dichotomous variable distinguishing those who feel very or fairly safe from those who feel a bit or very unsafe when 'walking alone in this area after dark'.

5. A trichotomous division of Acorn neighbourhoods into high-, medium-, and low-risk types based on victimization rates (see Hough and Mayhew, 1985, and Note 13).

6. L: whether or not the interviewer assessed the streets around a household to contain 'a lot' of litter. H: whether or not interviewers assessed local houses to be in a fair or good physical state. S: whether or not respondents claimed it would be easy for them to distinguish strangers from residents.

7. RS: whether or not respondents are satisfied 'with living in the area'. C: whether or not the respondent lives in a neighbourhood where 'people do things together and try to help each other'.

8. Young (16-30), middle-aged (31-60) and elderly (over 60).

9. Four categories: Women and men separately in (a) single-adult households and (b) multiple-adult households.

10. Owners and council renters.

11. $O_s = 1.2 \times 1.2T \times 1.5A_1 \times 1.2A_2 \times -2.4F_1 \times -2.3F_2 \times 2.1F_3 \times 1.3R_1 \times 1.1R_2 \times 1.1C \times 1.4RS \times 1.4S \times -1.2L \times 1.1H$

Where

O_s = the odds of feeling safe rather than unsafe and independent terms refer to the effect on O_s of:

T: owning rather than renting

A_1: being young rather than old;

A_2: being middle-aged rather than old;

F_1: being a single female rather than a 'family' male;

F_2: being a 'family' female rather than a 'family' male;

F_3: being a single male rather than a 'family' male;

R_1: being in a low- rather than a high-risk neighbourhood type;

R_2: being in a medium- rather than a high-risk neighbourhood type;

C: feeling a sense of community spirit;

RS: feeling satisfied with local life;

S: being able to distinguish strangers from residents;

L: living in a littered environment;

H: living in a neighbourhood where housing conditions are good.

The model fits well (likelihood ratio X^2 = 329.97, 2290 d.f., $p \rightarrow 1$) and provides a proportional reduction of errors (over the unpredictability of fear when the distribution of independent variables is unknown) of 14.6 per cent - i.e. index of concentration = 0.146 - a measure analogous to Goodman and Kruskal's Tau (see Blalock 1979, 307-10).

12. The term 'family' male or female is used to indicate that respondents live with a 'family' of one or more other adults (with or without children).

13. The incidence of victimization as measured by the 1984 BCS allowed the various sub-categories of the Acorn classification of enumeration districts to be grouped into high-, medium- and low-risk neighbourhood types. Hough and Mayhew (1985) show, for instance, that burglary rates vary from 1-3 per cent of households per year in low-risk areas to 10-12 per cent in high-risk neighbourhoods; the corresponding proportions for rates of robbery/theft from the person are 0.9-1.1 per cent persons per year to 3.3-4.3 per cent.

References

Agnew, R.S. (1985) 'Neutralizing the impact of crime', Criminal Justice and Behaviour, 12, 221-39

Baldwin, J. and Bottoms, A.E. (1976) The urban criminal, London: Tavistock

Baumer, T.L. (1978) 'Research on fear of crime in the United States' Victimology 3, 254-67

Baumer, T.L. (1985) 'Testing a general model for fear of crime: data from a national sample', Journal of Research in Crime and Delinquency 22, 239-55

Bentham, G. (1986) 'Socio-tenurial polarization in the United Kingdom 1953-83: the income evidence', Urban Studies 23, 157-62

Berg, W.E. and Johnson, R. (1979) 'Assessing the impact of victimization', in W.W. Parsonage (ed.) Perspectives on victimology, Beverly Hills and London: Sage 58-71

Blalock, H.M. (1979) Social Statistics (revised 2nd edn) New York: McGraw-Hill

Block, R. (1971) 'Fear of crime and fear of the police', Social Problems 19, 91-101

Bottoms, A.E. and Wiles, P. (1986) 'Housing tenure and residential community crime careers in Britain', in A.J. Reiss, Jr, and M. Tonry (eds) Communities and Crime, London and Chicago: University of Chicago Press, 101-62

Bottoms, A.E. and Wiles, P. (1987) 'Crime and housing policy: a framework for crime prevention analysis', in T. Hope and M. Shaw (eds) Communities and Crime Reduction, London: HMSO

Box, S., Hale, C. and Andrews, G. (forthcoming) 'Explaining fear of crime', British Journal of Criminology

Brantingham, P.J., Brantingham, P.L. and Butcher, D. (1986) 'Perceived and actual crime risks', in P.M. Figlio, S. Hakim, and G.F. Rengert (eds) Metropolitan Crime Patterns, New York: Criminal Justice Press, 139-60

Braungart, M.H., Hoyer, W.J. and Braungart, R.G. (1979) 'Fear of crime and the elderly', in A. Goldstein, W. Hoyer and P. Monti (eds) Police and the Elderly, New York: Pergamon Press, 15-29

Brooks, J. (1974) 'The fear of crime in the United States', Crime and Delinquency 20, 241-4

Clarke, A.H. and Lewis, M.J. (1982) 'Fear of crime among the elderly', British Journal of Criminology 22. 49-62

Clemente, P. and Kleiman, M.B. (1977) 'Fear of crime in the

United States', Social Forces 56, 519-31

Crook, A.D.H. (1986) 'Privatization of housing and the impact of the Conservative Government's initiatives on low-cost home ownership and private renting between 1979 and 1984 in England and Wales: 1. The privatization policies', Environment and Planning A, 18, 639-59

DuBow, F. and Emmons, D. (1981) 'The community hypothesis', in D.A. Lewis (ed.) Reactions to crime, Beverly Hills and London: Sage, 161-81

Duncan, J.S. (1981) Housing and Identity, London: Croom Helm

Evans, D.J. (1980) Geographical Perspectives on Juvenile Delinquency, Aldershot: Gower

Eve, R.A. and Eve, S.B. (1984) 'The effects of powerlessness, fear of social change, and social integration on fear of crime among the elderly', Victimology 9, 290-5

Feinberg, N. (1981) 'The emotional and behavioural consequences of violent crime on elderly victims', Victimology 6, 355-7

Forrest, R. and Murie, A. (1983) 'Residualization and council housing: aspects of the changing social relations of housing tenure', Journal of Social Policy 12, 453-68

Forrest, R. and Murie, A. (1986) 'Marginalization and subsidized individualism: the sale of council houses in the restructuring of the British welfare state', International Journal of Urban and Regional Research 10, 46-66

Forrest, R. and Murie, A. (1987) 'The pauperization of council housing', Roof, Jan.-Feb., 20-23

Garofalo, J. (1979) 'Victimization and the fear of crime', Journal of Research in Crime and Delinquency 16, 80-97

Garofalo, J. (1981a) 'The fear of crime and its consequences', Journal of Criminal Law and Criminology 72, 829-57

Garofalo, J. (1981b) 'Crime and the mass media: a selective review of research', Journal of Research in Crime and Delinquency 18, 319-50

Garofalo, J. and Laub, J. (1978) 'The fear of crime: broadening our perspective', Victimology, 242-53

Giles-Sims, A. (1984) 'A multivariate analysis of perceived likelihood of victimization and degree of worry about crime among older people', Victimology 9, 222-33

Henig, J. and Maxfield, W.G. (1978) 'Reducing fear of crime:

strategies for intervention', *Victimology* 3, 297-313

Herbert, D.T. (1982) *The Geography of Urban Crime*, London: Longman

Herbert, D.T. and Hyde, S. (1984) *Residential crime and the urban environment*, a report to the Economic and Social Research Council, University College Swansea, Department of Geography

Hindelang, M.J., Gottfredson, M.R. and Garofalo, J. (1978) *Victims of Personal Crime: an Empirical Foundation for a Theory of Personal Victimization*, Cambridge, Mass: Ballinger

Holdsworth, D.W. (1979) 'House and home in Vancouver', in G. Stelter and A. Artibise (eds) *The Canadian City: Essays in Urban History*, Toronto: Macmillan, 186-211

Hope, T. (1986a) 'Community and environment in recent criminology', in D. Canter and M. Krampen (eds) *Actions and Place*, Vol. 1, Aldershot: Gower

Hope, T. (1986b) 'Council tenants and crime', *Research Bulletin* 21, London: Home Office Research and Planning Unit

Hope, T. and Hough, M. (1987) 'Area crime and incivilities: a profile from the British Crime Survey', in T. Hope and M. Shaw (eds) *Communities and Crime Reduction*, London: HMSO

Hough, M. (1984) 'Residential burglary: a profile from the British Crime Survey', in R.V.G. Clarke and T. Hope (eds) *Coping with Burglary*, Boston: Kluwer-Nijhoff, 15-28

Hough, M. (1985) 'The impact of victimization: findings from the British Crime Survey', *Victimology* 10, 488-97

Hough, M. and Mayhew, P. (1983) *The British Crime Survey: First Report*, London: HMSO

Hough M. and Mayhew, P. (1985) *Taking Account of Crime: Key Findings from the British Crime Survey*, London: HMSO

House, J.W. (ed.) (1978) *The UK Space* (2nd edn), London: Weidenfeld and Nicolson

Hunter, A. and Baumer, T.L. (1982) 'Street traffic, social integration and fear of crime', *Sociological Inquiry* 52, 122-31

Jaycox, V.H. (1978) 'The elderly's fear of crime: rational or irrational' *Victimology* 3, 329-33

Kail, B.L. and Kleinman, P.H. (1985) 'Fear of crime, community organization, and limitations on daily routines', *Urban Affairs Quarterly* 20, 400-8

Katzman, M.T. (1980) 'The contribution of crime to urban decline', Urban Studies 17, 277-86

Kennedy, L.W. and Krahn, H. (1983) 'Rural-urban origins and fear of crime: the case for "rural baggage"', Edmonton Area Series Report 28 Department of Sociology, University of Alberta

Kennedy, L.W. and Silverman, R.A. (1985a) 'Significant others and fear of crime among the elderly', Journal of Aging and Human Development 20, 241-56

Kennedy, L.W. and Silverman, R.A. (1985b) 'Perceptions of social diversity and fear of crime', Environment and Behaviour 17, 275-96

Kerner, H.J. (1978) 'Fear of crime and attitudes towards crime: comparative criminological reflections', International Annals of Criminology 17, 83-102

Kinsey, R. (1984) Merseyside Crime Survey: First Report, Merseyside County Council

Krahn, H. and Kennedy, L.W. (1985) 'Producing personal safety: the effects of crime rates, police force size, and fear of crime', Criminology 23, 697-710

Lavrakas, P.J. and Herz, E.J (1982) 'Citizen participation in neighbourhood crime prevention', Criminology, 20, 479-98

Lavrakas, P.J. and Lewis, D.A. (1980) 'Conceptualization and measurement of citizens' crime prevention behaviours', Journal of Research in Crime and Delinquency 17, 254-73

Lewis, D.A. (1980) Sociological Theory and the Production of a Social Problem: the Case of Fear of Crime, Reactions to crime project, Chicago: Northwestern University, Center for Urban Affairs

Lewis, D.A. and Maxfield, M.G. (1980) 'Fear in the neighborhoods: an investigation of the impact of crime', Journal of Research in Crime and Delinquency 17, 140-59

Lewis, D.A. and Salem, G. (1981) 'Community crime prevention: an analysis of a developing strategy', Crime and Delinquency 27, 405-21

Lohman, J.S. (1983) 'Fear of crime as a policy problem', Victimology 8, 336-43

McPherson, M. (1975) 'Realities and perceptions of crime at the neighbourhood level', Victimology, 3, 319-28

Malpass, P. (1986) 'Low-income home ownership and housing policy', (review article), Housing Studies 1, 241-5

Malpass, P. and Murie, A. (1982) Housing Policy and

Practice, London: Macmillan

Markson, E. and Hess, B. (1980) 'Older women in the city', Signs 5, 127-41

Martin, R. and Rawthorn, B. (eds) (1986) The Geography of De-industralization, Basingstoke: Macmillan

Mawby, R.I. and Brown, J. (1984) 'Newspaper images of the victim: a British study', Victimology 9, 82-94

Maxfield, M.G. (1984a) Fear of Crime in England and Wales, London: HMSO

Maxfield, M.G. (1984b) 'The limits of vulnerability in explaining crime: a comparative neighbourhood analysis', Journal of Research in Crime and Delinquency 21, 233-50

Maxfield, M. (1987) Explaining fear of crime: Evidence from the 1984 British Crime Survey, Home Office Research Paper No. 41, London: Home Office

Mayhew, P. (1985) 'The effects of crime: victims, the public, and fear', in European Committee on Crime Problems, Research on Victimization, (Collected Studies in Criminological Research, 23), Strasbourg: Council of Europe, 67-103

Merry, S.E. (1981) Urban Danger, Philadelphia: Temple University Press

Miethe, T. and Lee, G.R. (1984) 'Fear of crime among older people: a reassessment of the predictive power of crime-related factors', The Sociological Quarterly 25, 397-415

Mugford, S. (1984) 'Fear of crime - rational or not? A discussion and some Australian data', The Australian and New Zealand Journal of Criminology 17, 267-75

NOP (1985) British Crime Survey Technical Report, London: NOP

Normoyle, J. and Lavrakas, P.J. (1984) 'Fear of crime in elderly women: perceptions of control, predictability and territoriality', Personality and Social Psychology Bulletin, 10, 191-202

Patterson, A.H. (1978) 'Territorial behaviour and fear of crime in the elderly', Environmental Behaviour and Non-verbal Behaviour 2, 131-44

Patterson, A.H. (1979) 'Training the elderly in mastery of the environment', in A. Goldstein, J. Hoyer and P.J. Motin (eds) Police and the Elderly, New York: Pergamon, 86-94

Pease, K. (forthcoming) Judgements of Crime Seriousness: Evidence from the 1984 British Crime Survey, Research and Planning Unit Paper No. 42, London: Home Office

Podolefsky, A.M. (1979) Reactions to Crime Papers (number/title unspecified), Chicago: Center for Urban Affairs, Northwestern University

Porteus, J.D. (1976) 'Home: the territorial core'. Geographical Review 66, 383-90

Reiss, A.J. Jr. (1986) 'Why are communities important in understanding crime?', in A.J. Reiss Jr. and M. Tonry (eds) Communities and Crime, London and Chicago: University of Chicago Press, 1-33

Reiss, A.J. Jr and Tonry, M. (eds) (1986) Communities and Crime, London and Chicago: University of Chicago Press

Riger, S. (1985) 'Crime as an environmental stressor', Journal of Community Psychology 13, 270-80

Riger, S. and Skogan, W.G. (1978) 'Introduction', Victimology 3, (special issue on the fear of crime)

Riger, S., Gordon. M.T. and LeBailly, R. (1978) 'Women's fear of crime: from blaming to restricting the victim', Victimology 3, 274-84

Riger, S., Gordon, M.T. and LeBailly, R. (1982) 'Coping with crime: women's use of precautionary behaviours', American Journal of Community Psychology 10, 369-86

Robinson, R. (1986) 'Restructuring the welfare state: an analysis of public expenditure 1979/80-1984/5', Journal of Social Policy 15, 1-21

Sampson, R.J. and Wooldridge J.D. (1986) 'Evidence that high crime rates encourage migration away from central cities', Sociology and Social Research 70, 310-14

Saunders, P. (1984) 'Beyond housing classes: the sociological significance of private property rights in means of consumption'. International Journal of Urban and Regional Research 8, 202-7

Saunders, P. (1986) Social Theory and the Urban Question (2nd edn), London: Hutchinson

Shotland, R., Hayward, S.. Young, D., Signoralla, M., Mindingall, K., Kennedy, J., Rovine, M. and Danowitz, E. (1979) 'Fear of crime in residential communities', Criminology 17, 34-45

Silverman, R.A. and Kennedy, L.W. (1984) 'Loneliness, satisfaction and fear of crime: test for non-recursive effects', Canadian Journal of Criminology 27, 1-13

Skogan, W. (1986) 'Fear of crime and neighbourhood change', in A.J. Reiss, Jr and M. Tonry (eds) Communities and

Crime, London and Chicago: University of Chicago Press, 203-29

Skogan, W.G. and Maxfield, M.G. (1981) Coping with Crime, Beverly Hills and London: Sage

Smith, D.J. and Gray, J. (1985) Police and People in London, Aldershot: Gower

Smith, S.J. (1983) 'Public policy and the effects of crime in the inner city', Urban Studies 20, 229-39

Smith. S.J. (1985) 'News and dissemination of fear', in J. Burgess and J. Gold (eds) Geography, the Media and Popular Culture, London: Croom Helm, 229-53

Smith, S.J. (1986) Crime, Space and Society, Cambridge: Cambridge University Press

Smith, S.J. (1987) 'Fear of crime: beyond a geography of deviance'. Progress in Human Geography 11, 1-23

Sparks, R.F., Genn, H. and Dodd, D.J. (1977) Surveying Victims, Chichester: Wiley

Stafford, M.C. and Galle, O.R. (1984) 'Victimization rates, exposure to risk, and fear of crime', Criminology 22, 173-85

Sundeen, R.A. and Matthieu, J.T. (1976) 'The fear of crime and its consequences among elderly in three urban communities', The Gerontologist 16, 211-19

Taub, R.P., Taylor, D.G. and Dunham, J.D. (1981) 'Neighbourhoods and Safety' in D.A. Lewis (ed.) Reactions to Crime, Beverly Hills and London: Sage, 103-19

Taub, R.P., Taylor, D.G. and Dunham, J.D. (1984) Paths of Neighbourhood Change, Chicago: University of Chicago Press

Taylor. R.B., Gottfredson. S.D. and Brower, S. (1984) 'Block crime and fear: defensible space, local social ties, and territorial functioning', Journal of Research in Crime and Delinquency 21, 303-31

Taylor, D.G., Taub, R.P. and Peterson, B.L. (1986) 'Crime, community organization, and causes of neighbourhood decline', in R.M. Figlio, S.Hakim, and G.F. Rengert (eds) Metropolitan Crime Patterns, New York: Criminal Justice Press, 161-77

Warr, M. (1980) 'The accuracy of public beliefs about crime', Social Forces 59, 456-70

Warr, M. (1982) 'The accuracy of public beliefs about crime: further evidence', Criminology 20, 185-204

Warr, M. (1984) 'Fear of victimization: why are women and the elderly more afraid?', Social Science Quarterly 65,

691-702

Warr, M. (1985) 'Fear of rape among urban women', Social Problems 32, 238-50

Wilson, J.Q. and Kelling, G.L. (1982) 'Broken windows: the police and neighbourhood safety', The Atlantic Monthly, March, 29-38

Wood, D.S. (1984) British Crime Survey: Technical Report, London: Social and Community Planning

Worrall, A. and Pease, K. (1986) 'Personal crime against women: evidence from the 1982 BCS' The Howard Journal 25, 118-24

Young, J., Jones, T. and Maclean, B. (1986) The Islington Crime Survey, Aldershot: Gower

Chapter Eleven

THE GEOGRAPHY OF SOCIAL CONTROL: CLARIFYING SOME THEMES

John Lowman

Throughout its history the geography of crime has focused almost exclusively on the milieu of the criminal actor or the nature of the criminal event in isolation from the control system which defines and processes 'deviance'. Geographers interested in crime have generally followed the mandates of traditional criminology (i.e. 'positivist', 'functionalist', and 'criminal justice' perspectives) by treating 'crime' and 'justice', or 'deviance' and 'control' as analytically discrete domains (Lowman 1982, 1983, 1986a). Generally, the geography of crime has not reflexively engaged in the universe of meaning and action which it is about. In contrast, what are variously termed 'interactionist', 'ethnomethodological', 'phenomenological', and 'subjectivist' brands of criminology (the 'sociology of deviance') and various types of 'critical', 'radical', 'Marxist' and 'socialist' perspectives reason that crime cannot be adequately theorized in isolation from the context of its definition and control; ultimately, deviance and social control are analytically inseparable. According to this logic the geography of crime must, at least partly, be reconceptualized as a geography of social control. While the advent of a 'left realism' in criminology (cf. Kinsey, Lea, and Young 1986; Lea and Young 1984; Rafter 1986; Young 1986) has heralded a renewed critical theoretic interest in 'street' and 'lower-class' crime (as opposed to the crimes of the powerful, and arcane crime more generally) the analysis of social control institutions - particularly the role and activity of the police in relation to both the formation of criminogenic sub-cultures and to community crime-reporting behaviour - remains central to the left realist perspective.

But the concept of 'social control' and its role in the development of social geography generally and the geography of crime in particular require considerable elaboration, especially given Harries's (1986; 578) misunderstanding of this concept. It is to the task of clarifying and qualifying a geographical perspective on social control that this paper is devoted.

The paper is divided into four parts. The first section provides a short history of the social control concept; the second section examines recent literature pertinent to the development of a geography of social control; the third suggests that the introduction of the social control concept to the geography of crime should not be read as a simple inversion of traditional criminology; the final section of the paper, a reaction to recent criticisms of the nascent geography of social control, reasserts the importance of the social control concept in developing a critical geographic perspective on crime.

The social control concept

Like most over-arching concepts in sociology, the term 'social control' has appeared in a variety of guises. Unfortunately, there has been a tendency for so broad an application that virtually any social process can be considered to be an instrument of social control. The result of this over-abundance of meaning is two-fold. Some authors have concluded that the concept, now devoid of any analytic utility, should be abandoned (Chunn and Gavigan 1986); others have concluded that its use, being unavoidable, should be strategic - it should be employed for a particular theoretical or heuristic purpose. My use of the concept is strategic; it is designed to draw attention to a particular weakness in the geography of crime. But others may wish to use a different strategy in employing the concept in geography. To this extent, an excursion through the wider literature on social control is provided in this paper.

While almost every sociological grand theorist has attempted to explain general mechanisms of social order, there are four main senses (often interwoven) in which the formal term 'social control' has been used: as a central concept of classical sociological grand theory; as a social psychological mechanism; as the institutionalized reaction (both formal and informal) to perceived deviance; and, in a return to grand theory, as an instrument of power. (1)

The puzzle of order: social control in grand theory

In its most fundamental sense, the term 'social control' referred to 'the capacity of a society to regulate itself according to desired principles or values' (Janowitz 1975, 82). In this sense, even though not used explicitly until the work of the American E.A. Ross (e.g. 1901), the concept was implicated in the very question that originally gave sociology its meaning: what makes social order possible? Thus, while Durkheim's functionalist perspective did not actually use the concept as such, his general concern with moral order and his search for the 'determination of moral facts' was very much a concern with what later writers designated as 'social control' (Janowitz 1975, 91).

In Durkheim's writing, social stability resulted from effective social organization (Meier 1982, 36), with social control mechanisms envisaged as emergent phenomena, beyond the workings of human agency and political ideology. Social control became one of the central concepts in sociology, particularly as an explanation of the maintenance of social order (or conversely its breakdown) in highly complex societies during periods of their transition from one form to another. It was particularly the attempt to understand the transition from rural to urban society, and to explain how order was possible in urban societies where humans were freed from the constraints of the primary group, that gave the formal social control concept its initial impetus. It was this conception that came to dominate early American sociology.

Social control in early American sociology: social psychological perspectives

Given the broad sociological interest in social control and social change that surfaced in the formative years of European sociology, the United States provided the quintessential setting for the study of social change (Meier 1982, 37), particularly in terms of the contrasting control mechanisms characterizing gemeinschaft and gesellschaft societies (Tonnies 1887). Part of the original impetus towards the development of a specific social control concept was in response to Tonnies's model, which was judged to be over-simplified and inadequate for theorizing social change (Janowitz 1975, 85).

Ross's (1901) conception of social control must be understood as a reaction to the model of human nature prevailing in economics, the discipline in which he was trained (Meier 1982, 37). Economists portrayed humans as inherently self-serving. In contrast Ross's concern was with the intersection of collective and self-interests, and with how to maximize collective interests in urban societies where the social bonding mechanisms of primary group interactions no longer held sway.

These same types of concerns were carried over into the works of Chicago school social ecologists (such as W.I. Thomas, Robert Park, Ernest Burgess, and Clifford Shaw), the social behaviourism of George Herbert Mead, and into philosophical pragmatism. (2)

For Park and Burgess the concept of social control was at the very heart of the sociological enterprise, for, as they put it in their Introduction to the Science of Sociology, 'all social problems turn out to be problems of social control' (Park 1921, 785). In the writings of the Chicago sociologists - particularly important because of their place in the intellectual heritage of the geography of crime - the question was how to produce surrogates for primary group social control mechanisms so as to maximize 'social' control and minimize the need for 'coercive' control, i.e. the use of legitimate force to achieve compliance (c.f. MacIver and Page 1949).

Conceived of in this way, social organization resulted from effective social control; disorganization was the product of a breakdown of control mechanisms. This formula was prohibitively circular. Social disorganization was defined as the absence of social control; social control was identified by the absence of disorganization - social organization was social control (Meier 1982, 39, 41). Also problematic was the normative soul of the theory - the concept of social disorganization was a moral invective reflecting the values of the theorist rather than the objective 'scientific' concept that its architects claimed it to be (Mills 1943; Schwendinger and Schwendinger 1974). It failed to acknowledge the sort of argument made subsequently in the sociology of deviance that inner-city communities and deviant groups were often highly organized, a point nicely made by Whyte's (1943) article strategically entitled, 'Social organization in the slums'. Thus it was not necessarily that the norms of rural or old world societies broke down in urban settings through a process of disorganization, but that different norms developed. Shaw and McKay's perspective also failed to provide a political economy of North American urban life and the immigrant's place within it. From a critical perspective social disorganization could be reinterpreted as political economic marginalization. Recent advocates of such a theory also problematize the pluralism of the sociology of deviance by arguing that while certain outlaw sub-cultures might conform to value systems that are distinctly their own, these value systems must nevertheless be understood as often deriving from and very much in line with the consumerist principles of contemporary capitalist societies (cf. Lea and Young 1984, ch. 3).

It is, nevertheless, the social disorganization conception of social control that appears to have provided the implicit foundation of much of the geography of crime. In this kind of perspective control institutions, because they are treated as simply reactive to deviance, are not taken into account in the discussion of crime. Crimes of the powerful have been ignored altogether; perhaps it is because they do not serve the disciplinary needs of a geography linking physical proximity and neighbourhood traits to crime causation. (3)

Other disciplines

Developments in the social control concept were not confined to sociology. In economics, concern has been shown for the mechanisms outside the competitive economic process which influenced resource allocation, especially in terms of co-operative arrangements supplied by legislative mechanisms - in this way the state became directly implicated in the process of social control. The conception of the state's role was as architect of social good according to some consensual blueprint - there was little indication that, at least in some instances, the state served as an instrument to secure particular interests. Such was the case with Roscoe Pound's portrayal of the state in his Social Control Through Law (1942). In Pound's view, as the influence of custom and kinship weakened, other mechanisms - particularly law - gained greater importance. For Pound, law was the rationally imposed authority of the state.

Pound's perspective was generally in keeping with the

influential functionalist view of social control originating in Durkheim's work. Similar perspectives were developed by the sociologist Talcott Parsons (1951) and the anthropologist A.R. Radcliffe-Brown (1933). Social control for these theorists consisted of formal sanctions emerging out of official group sentiments, and informal sanctions operating at the level of interpersonal relations (especially in terms of the norms associated with the roles conferred by membership in various types of groups). Again, the main criticism of these perspectives relates to their assumption that the objects of social control, especially the formal variety, emerge from a social consensus about what norms should obtain. There is little sense of historical struggle over the form and content of law.

This macro-level analysis found its counterpart in a social psychology focusing on primary and secondary socialization processes. Research on social control in a variety of institutional settings proliferated after the Second World War, stimulated particularly by the appearance of Hughes's essay on 'Institutions' (Janowitz 1976, 97). The archetypal criminological version of this focus on socialization processes appeared in Travis Hirschi's 'control theory' (4), an explanation of normative compliance in terms of a person's 'belief' in, and 'attachment', 'commitment' and 'involvement' with the prevailing social order. What is missing from his account is a systematic analysis of the factors responsible for the level of a person's attachment, commitment, involvement, and belief. And like the concept of social disorganization before it, this attempt to explain deviance in terms of the failure of social control runs 'the risk of "explaining" something (deviance) by mere definition (the absence of social control)' (Meier 1982, 46). Again in Hirschi's perspective the content of normative order remains unquestioned.

Social control as a reaction to perceived deviance and as an instrument of power: pluralist and class conflict models

As well as its emphasis on the invoking of rules, a subjectivist analysis of deviance indicated that rule breaking was often, paradoxically, norm-conforming behaviour. (5) Rendered this way, traditional questions about the differences between criminals and non-criminals were moot. If there was anything to be 'explained' it was the deployment of the labels 'mad' and 'bad', and the operation of the internal rules of deviant sub-cultures that mattered, not the presence or absence of distinct generic deviant (or, more specifically, criminal) dispositions, whether these be social or psychological in origin. The questions of crime causation and crime control were largely banished from the analysis.

Because subjectivist perspectives originally gained much of their identity and momentum from their opposition to traditional criminological inquiry, they tended to simply invert it by becoming much more concerned with the criminal labelling process than with crime. Thus the sociology of deviance in the 1960s became preoccupied with the development of a plurality of value systems in urban capitalist society, and the designation of some of those values as deviant.

Instead of the consensus model of law formation that underlay traditional perspectives in criminology the sociology of deviance posits a pluralist model of society and law. From this perspective the definition and application of the law assumed paramount importance. The transactions between deviants (mostly delinquents, 'morality' or 'victimless' crime offenders and mental health patients) and controllers (police officers, probation officers, psychiatrists), especially in terms of the controller's selection of deviants for official processing, gained a central place in the analysis.

The pluralist conception of law and social norms underlying the sociology of deviance found its counterpart in the power conflict perspectives of Vold (1958) and Turk (1969) (6) and in the class-conflict models of Marx<u>ian</u> (if not actually Marx<u>ist</u>) 'radical' criminology (for review see Lynch and Groves 1986). Power conflict theory, (inspired mainly by the work of Weber and Dahrendorf) held that law was an instrument used by dominant social groups to reproduce their power, but without explaining why some groups were able to gain ascendance. The first wave of Marxian criminology, left functionalism - usually referred to as 'instrumentalism' (7) or 'left idealism' (cf. Young 1979) - reinterpreted power struggles as <u>class</u> struggles, viewing law and the state as instruments of class domination. In this view, crime is an epiphenomenon of capitalism, and will only disappear once the need for the state and law has withered away. Other versions of socialist criminology - notably, what Young (1979) refers to as left reformism - have recognized that because state power is semi-autonomous (i.e. that while the state may serve the interests of capital accumulation, it also serves general interests, and acts in its own interests) social change can be affected through social reform. Left realism problematizes the distinction between consensus and conflict perspectives by recognizing that while conflicts over norms do exist, there is a consensus over the desirability of a core group of crimes (and to this one might add the observation that what is often more important than conflict is the social construction of consensus). What is common to all these perspectives is a direct concern with the actions of the state and other institutions in mobilizing resources against anticipated deviance. It is the relevance of this sense of social control to the geography of crime that is discussed here.

A geography of social control

The most important aspect of the literature on reactions to perceived deviance is that there is a sense in which control is constitutive of deviance. Such an insight demands a considerable conceptual remapping of any generic geographical perspective on crime. Earlier work (Lowman, 1982, 1983, 1986a) has attempted to indicate the variety of ways - from the conception of what constitutes the appropriate object of study, to the explanation of actual offender behaviour - in which control can be said to constitute deviance. This is not to argue that deviance must be understood solely in terms of control (a point to be returned to below) or that it is not appropriate to suspend an analysis of control in order to understand the behaviour of persons labelled deviant, or to minimize the need for a geographic victimology (cf. Smith 1986). Nor is it necessary to construct a geography of social control solely for the purpose of understanding deviance - social control is important in its own right.

A strategic definition

In his seminal *Visions of Social Control* (1985), Stanley Cohen defines social control as:

> those organized responses to crime, delinquency and allied forms of deviant and/or socially problematic behaviour which are actually conceived of as such, whether in the reactive sense (after the putative act

> has taken place or the actor been identified) or in the proactive sense (to prevent the act). These responses may be sponsored directly by the state or by more autonomous professional agents in, say, social work and psychiatry. Their goals might be as specific as individual punishment and treatment or as diffuse as 'crime prevention', 'public health', and 'community mental health'.
>
> (1985, 3)

Following this revisionist definition, social control refers to either the anticipation of, or the reaction to perceived deviance. The definition could be extended to include all planned intervention related to crime, mental health, and welfare; it could include both public (i.e. state-initiated) and private justice (in the sense described by Henry, 1983, 1987) (8) and civil law. (9) At the heart of the approach is a concern with power.

In this paper my primary interest is in the geography of planned intervention, and in the expression of different types of power, whether conceptualized in Marxian, Weberian, or Foucaultian terms. (10) At issue is the power to make law, the power to enforce it, the power to avoid criminal prosecution, the power to avoid victimization, and the power underlying our notions of what constitutes wrongdoing in the first place. Examples of the kind of analysis envisaged here can be found throughout the sociology of social control. For illustrative purposes I examine here the work of Michel Foucault and Stanley Cohen, analyses which have already been suggested (Lowman 1986a, 90) as providing useful signposts for a geography of social control. This return to Foucault's work is designed to show that it is not really about 'corrections' - as Harries's (1986, 578) critique of the nascent geography of social control suggested. (11) Rather it is about the decentralization of power that the transition from corporal to carceral forms of punishment and social control betokens.

It is also a period of putative decentralization of control that Stanley Cohen discusses in Visions of Social Control (1985, see also Cohen 1987). For the purposes of the present discussion I focus on his categorization of social control into 'exclusionary' and 'inclusionary' techniques.

Foucault's geography of panoptic power

In Discipline and Punish Michel Foucault (1977) begins to articulate a theory of power radically different from liberal juridical and Marxist conceptions. Foucault's analysis of punishment, particularly his discussion of Bentham's 'panopticon' (a prison design), is a vehicle for describing a transformation of power that occurred during the seventeenth, eighteenth and nineteenth centuries in western societies. The panoptic principles of discipline, surveillance, hierarchy, and classification represent to Foucault a type of power that is to be found in many social forms, not just the prison. It is as much a part of the asylum, the school, the army barracks, the clinical hospital, the factory - indeed the whole of the nineteenth- and twentieth-century city (12) - as it is a part of the penitentiary regime.

The transition from corporal to carceral punishment marked the transformation of the centralized power of a sovereign (when every crime was a 'regicide' of sorts) to a form of radically decentralized disciplinary power dispersed throughout the social fabric. Corporal punishment was directed against the body in an unrefined system of power that gave very little opportunity for subtle gradation. Power was centralized, emanating from the sovereign; punishment was a matter of vengeance. The system of decentralized power that took its place - exemplified by the panoptic principles of discipline, surveillance, and hierarchicalized classification - inserted the power to punish much more deeply into the social body. Foucault argues that this change resulted from the general inefficacy and socially disruptive nature of corporal punishment, not its inhumanity (cf. Cousins and Hussain 1984, ch. 7); setting a limit to punishment was probably not as important to reformers as the desire to punish even the least crime. So it was that the new system of power created a plethora of new delinquencies (whether legally proscribed or not). In the carceral system, punishment was not so much a matter of 'correction' (which implies returning something to a state that it had erred from), as it was a matter of moulding the person, of constituting normality and deviance.

By examining punishment as a form of power, Foucault is able to show how the same kind of principles providing the logic of the panopticon come to be exercised throughout society. In accordance with this system of power a new system of organizing social and physical space developed. A new political anatomy developed (an 'anatomy of detail') in which the partitioning of physical and social space became a tactic of control (see also Cohen 1985, ch. 6). The art of distributions which came to the fore in the control of plague and leprosy, was carried over into regulation of armies, and gradually into education, medicine, economy, factories, etc. Administrative and political domains became therapeutic spaces, while 'dangerous mixtures' - crime and disease - were partitioned off. Foucault's use of the term 'carceral' refers not to incarceration as such, but to techniques of hierarchical surveillance, coercion by means of observation, and the disciplinary enclosure of various spaces. These 'projects of docility' were not new to the seventeenth and eighteenth centuries; what was new was their combination at this time, a veritable 'swarming of disciplines' in a period when people were individualized, spaces turned into distinct singularities, and domination became less visible.

Though not conceived of as 'geography', Discipline and Punish is very much a geography of power. Panoptic power is a technique, one which provides the logic of a variety of social-geographic forms (see South 1987). For Foucault, power is therefore not simply something that is exercised by people - he is concerned with the constitutive or 'positive' nature of power (and what he terms 'power-knowledge' (13)). Consequently, his analysis is quite different from other histories of social control and the prison whether Marxian (e.g. Ignatieff 1978; Melossi and Pavarini 1981; Rusche and Kirchheimer 1939 (14)) or not (Rothman 1971), and is not easily reconciled with more familiar forms of Marxian critical theory. (15) But it does describe the geographic properties of a certain kind of power that came to the fore in the nineteenth century, and is still operative in contemporary capitalist (and socialist) societies, albeit in somewhat different forms. It is the analysis of these more recent social control trends that is taken up by Stanley Cohen.

Exclusion and inclusion

In Visions of Social Control (1985, 220-9) Stanley Cohen notes that methods of punishing and treating criminals are characterized by two spatial strategies: exclusion and inclusion. Exclusion refers to various systems of separation, segregation, isolation, expulsion, banishment, confinement and physical stigmatization. The most extreme form of exclusion is capital punishment, but the quintessential form in contemporary society is incarceration. Inclusion encompasses various systems of absorption, integration, and assimilation. 'Community corrections' and 'diversion' programmes represent contemporary styles of inclusion. While both exclusionary and inclusionary forms appear in a variety of ways in different societies:

> No iron law of political, economic or historical inevitability has determined which of these alternatives have been chosen at any particular time, nor is there any technique for making or predicting future choices.
> (266)

Visions of Social Control describes the results of the 'destructuring' movements (particularly decarceration, deinstitutionalization, community corrections, and diversion) which gained momentum during the 1950s, 1960s and 1970s in Europe and North America. The decarceration of asylums (Scull 1977) (16) and the advent of community corrections and diversion programmes apparently marked a renewed faith in inclusionary techniques. Although diversion programmes and community corrections promised to decrease control radically, Cohen (and a number of other authors) show that these movements led to an increase in control; rather than diverting people out of the system, they diverted them into it. Such programmes became supplements to incarceration, not alternatives. In a geographic sense, the proliferation of inclusionary controls creates an extension of disciplinary techniques both into the community and into private space, a further extension of what Foucault terms the carceral archipelago. This extension is supplemented by the techniques of a new behaviourism (17) (as represented by the development of situational crime-prevention theory, and behavioural approaches in the geography of crime). Modern inclusionary social control becomes a system of 'bleepers, screens and trackers', part of the 'invisibly controlling city' (Cohen 1985, 230).

Ironically the 'wider, stronger, and different nets' of inclusive responses to deviance are translated into further exclusionary practices. Rather than 'the community' becoming therapeutically regenerative, municipal politics, zoning regulations and neighbourhood interest groups combine to employ highly localistic, repressively parochial, and xenophobic strategies to carve out territory. The politics of exclusion 'with its metaphors of banishment, isolation and separation, its apparatus of walls, reservations and barriers, its utopia of the visibly purified city' (1985, 230) produces a moral urban landscape in which deviants are spatially and socially contained in ghettoes of neglect and zones of permissiveness.

In the case of mental patients the emptying of mental hospitals was not accompanied by the development of alternative treatment facilities in the community, so that community care amounted to little more than benign neglect (Scull 1977) steadily fuelling the ranks of the homeless and social worker case loads. More and more the mentally handicapped (deprived of their 'right to treatment') circulate through the soft end of the criminal justice and mental health systems as the dependent, defective, deranged and dangerous merge into an inclusive deviant category.

The overall effect of the increasing penetration of mechanisms of discipline and surveillance is that rather than community control, it is the control of communities (i.e. whole groups, populations, and environments) that is achieved (Cohen 1985, 127).

Against the inversion of traditional criminology

From the point of developing a geography of social control, Cohen's work identifies the essential geographic core of social control strategies as the constant interplay between exclusionary and inclusionary tactics. His examination of the development of community corrections shows how attempts ostensibly designed to reduce control ironically produced an overall system expansion (Cohen 1985, 45). It should be noted that this expansion is particularly well documented in the case of juvenile deviants, but the evidence for these effects is not so readily available in the case of adults, and more empirical work on the relationship between crime rates, types of crime, and control expansion is required. For, as Cohen observes, 'it would be a bizarre type of theory that completely ignored the possibility that the expansion of the crime-control system over the past two decades ... is a direct response to increasing official crime rates'. Changes in control 'cannot explain away the real structural factors that would lead one to an increased crime rate during this period' (1985, 91). It is in this sense that a geography of control ought not to be interpreted as a simple inversion of traditional criminology.

Rejecting 'controlology': deviance and control as interdependent variables

What has variously been termed a prima facie, 'positivist' or 'realist' (18) interpretation of official crime statistics treats them as dependent variables, dependent in the sense that they index (albeit incompletely) the amount of crime that occurs. The 'institutionalist' perspective in contrast treats them as independent variables, independent in the sense that the amount of crime is actually a reflection of the amount of control that is exercised (for discussion of the competing arguments see Biderman and Reiss 1967; Bottomley and Coleman 1981; Smith 1986, 38-42). In the geography of crime a prima facie interpretation prevails. But any problematization of geographers' use of crime statistics (Lowman 1982, 1983, 1986a) should not be taken as a recommendation for a 'controlological' perspective (Ditton 1979).

Arguing that any talk of 'bias' in official crime statistics comprises a thoroughbred institutionalism, Jason Ditton advances a controlological interpretation. In this 'juristic' view, 'criminal activity' is the activity of calling activities crimes' (1979, 23). Following Kitsuse and Cicourel's (1963, 247-8) recommendation that official statistics can only be used to analyse the people that they were collected by rather than those they were collected of, Ditton argues that 'variations in official crime rates are allowable as evidence of control-waves: but never of crime-waves' (p. 24).

This reversal of the conventional assignation of dependent and independent variables uses the same positivist epistemology as the prima facie perspective which it is designed to replace - it simply changes the nature of correspondence between observational and theoretical terms (cf. Gregory 1978, 56-9) from offenders to controllers. Such a tactic should be avoided because of its explicit rejection of the possibility that control responds to the activities of legal subjects. The conception of control as an independent variable is insensitive to the reflexive character of crime and control. Moreover, the use of crime statistics to index control is itself thoroughly problematic.

Generally, crime rates are a very poor measure of

'control'. Ditton tries to reject any notion of a dark figure of control: 'the "real" crime rate is the reaction rate, and cannot logically be any greater or lesser than it is' (p. 21). But in this formula, Ditton is unable to account for the great range of police control activity that is never recorded in official statistics. Observational studies of police indicate that very little time is spent processing criminals. Often, to achieve order in the streets the police do not have to use the criminal law at all; their presence alone is sufficient (Ericson 1982, 38). More problematic still is the inference that control is only exercised by police. This view ignores the work of a whole host of other professionals and para-professionals and the vital control function that social workers perform; it pushes psychiatry outside the arena of control altogether; and it ignores private security and many kinds of private justice. (19)

Most problematically of all, changes in public crime-reporting rates (by far the most important source of crime reports) are explained away by Ditton in a few lines as 'fantasy rises' (p. 13) which 'occur when feelings of moral entrepreneurship infuse a greater degree of attentiveness in the deviant audience (sic)'. Such an explanation (for which he offers virtually no empirical support) (20) would seem hard put to account for the enormous increases in reported crime rates in North America since 1960 (cf. Lowman and Menzies 1987, 113) and would certainly seem to be inadequate for understanding the regular seasonal fluctuations that characterize many crime rates (cf. Harries 1980, 106-12) especially crimes like residential burglary which are almost always discovered by members of the public. All in all, controlology offers a very poor understanding of control.

Generally then, crime statistics should not be conceptualized in terms of the piebald world of statistical instrumentalism; crime and control are best viewed as interdependent variables.

Left realism: beyond left idealism and the sociology of deviance

A growing number of theorists have recently suggested that the (over)emphasis on deviance definition carried over from the sociology of deviance into the critical literature has relegated populist concerns about crime to the 'realists' of the new right. To counteract this tendency a left realism evinces a new interest in crime, the victims of crime, and crime control (e.g. Lea and Young 1984; Kinsey, Lea, and Young 1986; Rafter 1986; Taylor 1981). Jock Young claims that the sociology of deviance and, in its turn, 'left idealism' (cf. Young 1979; see also Smith 1986 20-3) contained a number of unfortunate features which tended to minimize the consequences of criminal victimization. In particular, left idealism: tended to romanticize the criminal; ignored the victim of intra-working-class crime by drawing attention to the crimes of the powerful; viewed all crime as a product of class conflict; tended to explain why the state criminalizes certain people rather than explaining why certain people become criminal; ruled out the possibility of progressive reform; and argued that in a socialist political order the need for a criminal justice system (along with the state itself) would disappear (Lea and Young 1984, 95-104). In opposition to this view, left realism reasserts the seriousness of street crime, acknowledges that consensus does prevail over a core group of crimes, maintains that serious crimes ought to be prohibited, and advocates a variety of crime-control policy reforms, particularly in relation to policing. In short then, left realism counteracts any reluctance to engage in the problems of crime and crime control while still acknowledging that an analysis of capitalist social relations is essential to the explanation of 'marginal' criminality (a term used by Lea and Young 1984, 45-9) and crime victimization.

While all these criticisms of left idealism and left functionalism are well taken, (21) the swing of the pendulum heralded by the articulation of left realism ought not to be read as minimizing the need for an analysis of social control. As Lea and Young (1984, 62) in their description of left realism argue, (22) a synthesis of the various perspectives on crime and social control needs to occur - they should not be forced into a binary opposition. Thus one of the principal components of left realist perspectives on crime in Britain is an analysis of the effect of the change from 'consensus' to 'military' styles of policing the inner city on residents' perceptions of the police, flows of information about crime to the police, and on the effectiveness of the police to do anything about marginal crime. The increasing reliance on hard styles of policing - especially the differential use of stop and search powers targeting certain communities, races, or age groups - is argued as alienating the very persons who have the most information about crime. Since the public is the single most important source of information about marginal crime and the identity of offenders, this alienation undermines police power to do anything about crime, which in turn erodes the confidence of marginal communities in the police, which in turn creates the justification for the extension of hard measures, which further alienates marginal communities, and so on (Kinsey, Lea, and Young 1986, ch. 2; Lea and Young 1984, ch. 5). From this description of a classic deviancy amplification spiral it is clear that even the left realist call to take crime victimization seriously involves a theoretical perspective which reserves a central place for an analysis of the impact of social control mechanisms on crime.

The geographic literature is in a rather peculiar position vis-à-vis these debates. While Susan Smith's (1986) critical geography of crime can be located in the left reformist/left realist genre (and is thus very much concerned with policing) the geography of crime has arrived at this juncture without the intervening development of a sociology of deviance or a critical perspective on social control. Since my purpose is to demonstrate that the understanding of crime cannot be achieved in isolation from the context of its control, a short detour into Harries's (1986) critique of previous attempts to do this (Lowman 1982, 1983, 1986a) is warranted.

Control as constitutive of deviance

A geographic perspective on the social control of women

Harries asks, 'Can the demonstration of social control in the context of prostitution be equated with a similar exercise for a serious street crime such as aggravated assault?' While an analysis of proactive law enforcement in the case of prostitution (e.g. Lowman 1986c) should not be generalized to reactive law enforcement (23) an analysis of control in the context of prostitution certainly is relevant to an analysis of violence.

Harries asserts that prostitution is 'a "crime" that is exceeded in pettiness only by such activities as being the customer of a prostitute, being drunk in public, and playing hooky from school' according to a US Survey of Crime Severity. (24) From this observation Harries goes on to suggest that insights into the social control of prostitution are 'petty' and unhelpful for the analysis of a serious street crime such as aggravated assault.

A variety of problems are associated with these remarks, not the least of which involves the calibration of crime seriousness. Surely it is analytically significant that in 1978 in the United States almost 71 per cent of all first arrests of females were for prostitution offences, and that of the more than 89,000 persons arrested for prostitution only 10 per cent were (male) customers (Davis and Faith 1987). These statistics reveal that it is primarily over matters of sexuality that the criminal law is deployed against females in the United States. (25) But it is not so much to sexuality as it is to the public display of sexuality that law enforcement efforts are devoted; 85-90 per cent of the arrests of prostitutes are for offences in street locations. This distinction between public and private behaviour is an important one because it is vital to an understanding of the social control of women more generally. It is also important when it comes to understanding which women and girls become involved in street prostitution.

Generally, the distinction between public and private is thoroughly implicated in the whole process of deviance definition and social control and is fundamental to a geography of social control (cf. Kress 1980; Stinchcombe 1966). While the geography of crime has paid very little attention to gender issues (there is to my knowledge no geography of women's crime as such) a growing feminist literature in criminology, law, and sociology indicates that the distinction between public and private is crucial to the understanding of patriarchy, particularly in terms of the structure of male power over women, especially in the modern nuclear family. The 'coercion of privacy' (Stang Dahl and Snare 1978) must surely be taken into consideration when discussing the geography of 'aggravated assault'. Some of the most persistent forms of assault (including sexual assault) occur within the confines of the family. Some of these actions were not (and, in some countries, still are not) formally defined as criminal offences (it was not until 1983, for example, that a man in Canada could even be accused of raping his wife). The laws that do apply are generally under-enforced. Other types of domestic violence (usually perpetrated by males against females) are rarely reported to the police, and when they are, the tendency has been for each stage of the criminal justice system to resist the prosecution of offenders (as, ultimately, did many of the victims). Feminists charge that widespread and serious assault within the family has been tolerated in the name of family stability (thus the 'social problem' of 'sexual abuse' was not 'discovered' until the 1970s). It would seem that such assault is crucial to the general process of the defamilization of youth, and that it is lumpen youth who characteristically drift into street prostitution (Lowman 1987a).

Certainly, it would seem therefore, that an understanding of control deployment and public crime-reporting behaviour is relevant to the wider issue of how the meaning of the category 'assault' is socially constructed. Even if one wishes to restrict the discussion of crime to 'street' crime only, it should be remembered that street prostitutes usually begin their careers as juvenile runaways or throwaways; the reasons they report for leaving home usually involve serious assault (often including sex offences) by family members; (26) and once they become involved in street prostitution they are frequently the victims of rape, assault, and robbery (cf. Lowman 1984, ch. 6; Silbert and Pines 1981), and are frequently murdered in the course of their work. As victims of this form of common and serious street crime, prostitutes have often been treated differently from other victims, a reflection, at least in part, of the prostitute's status as a criminal. Perhaps crimes against them seem far less serious as a result (cf. Bland 1984). Unfortunately, Harries's suggestion that prostitution is 'petty' serves to reinforce such attitudes. If prostitutes were not subject to criminal censure, perhaps they would not be as vulnerable to victimization as they are in a place like Seattle, Washington, where some forty women (mostly street prostitutes) have fallen prey to the 'Green River killer' (Lowman 1986b). Cast in this light prostitution hardly seems to be a 'petty' matter at all.

Crimes of the powerful

Harries's more general comment that, 'Overall, the geography of social control would seem to offer little by way of new insight with respect to the crimes of violence responsible for so much fear and grief' (1986, 578) reveals a general ignorance of the burgeoning criminological literature on 'organizational', 'white-collar', and 'corporate' criminality. The first work to appear on white-collar crime (Sutherland 1945) was primarily designed to show that this type of infraction (whether formally defined as a crime or not) was just as serious as 'real' crime. A plethora of recent studies shows that crimes of the powerful certainly are responsible for a great deal of grief, although we may not fear them simply because we do not know when we are victimized by them (the perfect crime is one the victim is not aware of).

Thus public responses to a crime seriousness survey may be very poor indicators of what is 'serious'. How serious, for example, is tax evasion? It is quite likely that one of the most prevalent types of criminality (as measured by the number of offenders) is tax evasion. (27) We are currently experiencing in Britain and the US massive cutbacks in education, health and welfare, all in the name of fiscal restraint. But how should we evaluate the argument that government deficits are caused by welfare and education expenditures in the light of estimates that tax evasion in Canada and the US amounts to anywhere between 5 per cent and 15 per cent of Gross National Product (amounts that if collected would prevent further national debt) and against the counter-intuitive finding that the experience of a tax audit tends to increase the likelihood of a person's evasion of subsequent taxes (Brooks and Doob 1986)?

Obviously the arguments here are complex, but it does seem likely that the more funding is withdrawn from programmes aimed to ameliorate the conditions of the unemployed in western societies, the greater the levels of 'fear and grief' that Harries correctly attributes to street crime. Of course, the more that taxes are evaded (remembering that the regular wage earner is the person least able to evade taxes) the less the amount of money available in the public purse to finance all manner of social expenditures on health, welfare and unemployment. And what about the massive amounts of money that are controlled by the most economically consequential crimes of all - price fixing, illegal combines, and insider trading? It is almost impossible to open a current newspaper in Britain, Canada, or the US without finding stories about insider trading and stock market frauds of the most startling proportions - that make the dollar value of associated street crime pale by comparison (cf. Lea and Young 1984, 67) - and which must have profound implications for the functioning of market economies.

I can only assume from Harries's (1986) recent comments that his response to these observations would be that such infractions are not particularly important for an analysis of 'violence'. But a growing literature in criminology flatly disputes this kind of argument. Monahan, Novaco, and Geis (1979, 118), for example, define corporate violence as 'behaviour producing an unreasonable risk of physical harm to customers, employees, or other persons as a result of deliberate decision-making by corporate executives or culpable negligence on their part'. Using this kind of definition Ellis (1987, 94-7) estimates that the corporate violence rate in Canada is 28 times greater than the street violence rate, and the death rate more than 6 times greater than the 'street death rate'. The magnitude of these differences and overall validity of such comparisons is certainly open to debate. Often the harm caused by 'corporate violence' is much less directly intentional than the harm caused by theft and interpersonal violence. Street crime is transparent in its intentionality and immediate in its consequences (Lea and Young 1984, 65-75). An unknown percentage of the acts included by Ellis might not be defined as criminal. At this point the definition of what constitutes an offence becomes particularly tricky. How, for example, should we define the consequences of such actions as the American automobile industry's successful fifteen-year campaign against the legislation of compulsory installation of air bags in cars? Just how much 'fear and grief' has this carefully co-ordinated and thoroughly culpable campaign against saving drivers' lives actually caused?

These comments only just scratch the surface of the issues emerging in the rapidly expanding literature on crimes of the powerful. (28) What is conspicuously absent in the geographic literature is an analysis of corporate and white-collar crime and civil infractions of any sort, analyses which might yield new insights on the nature of such deviance and raise a variety of new questions. All sorts of new research areas are being opened up in criminology - there is growing interest in the conduct of lawyers, the practices of medical professionals are coming under more and more scrutiny (one wonders what an inter- and intranational geography of medical malpractice suits might demonstrate) etc. None of this is to suggest, for all the reasons articulated above, that street crime is unimportant. But it is to suggest that the geography of crime as it has been practised - and when the term crime is used generically - is ideologically distorting at worst, partial at best. It suggests that rather than using public opinion to direct criminological inquiry, theorists would do well to be sceptical about the dominant discourse on crime (perhaps more sceptical than left-realist writing might suggest). It would thus seem premature to dismiss the relevance of these new research directions to geographical criminology without the benefit of a careful programme of research.

Conclusions

The purpose of this paper has not been to present a fully fledged geography of social control, but to articulate the logical necessity of an analysis of control for a critical geography of crime. This is not to say that crime is explicable solely in terms of control, but that there are important areas in the study of crime, deviance, and control that have simply not been considered by geographers. And while left realism is emerging in criminology to articulate the shortcomings of a critical perspective which moved almost exclusively to a study of deviance definitions and the interests served by the formal and informal institutions of social control, the almost complete lack of a consideration of control variables in geographic perspectives on crime once again indicates the shortcomings of any approach that one-sidedly prioritizes crime.

But there is another and more fundamental critique that pervades these comments and that is an overriding suspicion of any perspective that attempts to confine itself to a single 'discipline'. In the case of the geography of crime this has led to a form of myopia that continues to encourage its practitioners to be overly selective in what they read, all in the name of producing analyses which are distinctively 'geographical'. In urging a wider reading of the extant literature on crime and control I have not attempted to recommend any one disciplinary perspective over others, or any single method (although my preference for 'critical' theory is clear enough). The literature I have urged geographers to consider is drawn from criminology, history, law, philosophy, political science, sociology, pyschology, and 'women's studies'. (29) Certainly there is virtue in synthesizing different literatures, and examining their implications for each other, especially given the sheer volume of contemporary academic writing. And given that a geography of crime exists (and may thus affect the world) it is certainly worth colonizing with new and critical ideas.

And not that there is anything wrong with the notion of 'disciplining' thought. But given that the best available geographies of social control (e.g. Foucault 1977; Cohen 1985) were neither written by geographers, nor conceived as such, one has to wonder just what is critical about the geography in a critical geography of crime and control - other than the professional interests of geographers in publishing papers.

What these comments do suggest, however, is that no matter what disciplinary classification structures our thought, social control, crime, and spatial structure ought to be theorized conjointly.

Notes

1. My discussion of the development of the concept is based on the commentaries of Cohen and Scull (1983), Janowitz (1975), Mayer (1983), Meier (1982), and Melossi (1987).

2. For a discussion of conceptions of social control in Chicago school sociology and American pragmatism more generally see Melossi (1987).

3. In contrast, the new behaviourism in geography eschews questions about causality altogether. Here the emphasis is on the administration of crime. The stress on the disembodied criminal act and the decontextualized environmental opportunity structure effectively rules out any analysis of the master institutions of society.

4. See also Walter Reckless's (1961) 'containment theory' and Albert Reiss's (1951) paper 'Delinquency as the failure of personal and social controls'.

5. The first alternatives to consensus perspectives in American criminology were provided by Edwin Sutherland's (1939) 'differential association theory' and various versions of sub-cultural theory (for a review see Brake, 1980). While these perspectives do not really contain a formal analysis of social control as such, they are important precursors to the sociology of deviance. Sub-cultural theory was imported into the geography of crime by David Herbert (1976, 1977) in work which also proclaimed the need for the application of the sociology of deviance to the geography of crime.

6. For a review of conflict theory more generally, see McDonald (1979).

7. This use of the term is confusing; in the philosophy of social science 'instrumentalism' has usually referred to a perspective allied to positivism. The main difference between the two is that positivism aims to identify causal laws, instrumentalism to predict events (see Gregory 1978; Keat and Urry 1975, Lowman 1986a).

8. For Henry (1987) private justice includes 'the practices of such institutions as the disciplinary bodies, boards, and councils of industrial and commercial organizations, professional and trade associations and unions, down to the peer sanctioning of relatively amorphous voluntary associations such as local self-help and mutual-aid groups'.

9. The concept could also be extended to encompass the more traditional sociological concern with customs and mores and other informal control mechanisms which induce compliance. But no matter how conceived, the study of social control should go far beyond Harries's (1986, 578) suggested focus on police, corrections, prosecutions, and justice.

10. This interest does not preclude a variety of other possibilities such as linking the study of social control to 'forensic geography' or to the analysis of law and landscape. But I will leave this task to other researchers.

11. Harries suggests that in my discussion of 'The spatial expression of control' (Lowman 1986a, 90) by invoking the work of Phil Cohen on policing and Foucault I appear 'to view "control" as a synonym for policing and corrections'. Even a cursory reading of Discipline and Punish would indicate that it is about much more than 'corrections'.

12. Images of the city have a special meaning for understanding visions of social control. The iconography of the modern western city involves various images of disorder: violence, crime, insecurity, pollution, and overcrowding. 'On the city streets' Cohen tells us 'lie the sharpest mirrors of dystopian imagery' (1985, 205). The solution to this chaos in the nineteenth century (mirrored in contemporary conceptions of defensible space) took the form of planning, regulation, classification, and surveillance. In nineteenth-century London streets were sometimes planned to cut through criminal 'rookeries' in the belief that physical redesign would solve social problems (Lowman 1982). The plans for the rebuilding of Vienna and Paris after the 1848 revolutions reflected the fear of urban insurrection and riot; wide streets and boulevards allowed a clear line for artillery

fire and troop movements (Hobsbawm, 1968, cited in Cohen 1985; 209). For a discussion of planning for order, see Cohen 1985, ch. 6).

13. For discussions of Foucault's analysis of power in Discipline and Punish see Cousins and Hussain (1984, Chs 7 and 9) Smart (1983, Ch. 4) and Poster (1984, Ch. 4). A selection of Foucault's writing and lectures on power and power/knowledge are to be found in Gordon (1980).

14. Again it should be pointed out (in contrast to Harries's misunderstanding of Foucault) that these analyses are similarly not about 'corrections', but about the wider social forces that mould strategies of punishment. They are examples of what Garland (1985) refers to as the 'social analysis of penality' (the analysis of the whole penal complex including sanctions, institutions, discourses, and representations) rather than 'penology' (which examines institutional practices simply in terms of the rhetoric supporting those practices).

15. For Foucault, power is not something exercised by human beings, and is not mimetic (i.e., is not reducible to political economy). Perhaps this means that his analysis is antithetical to the holistic conception of power developed by Marx and in various forms of Marxist or Marxian analysis; perhaps it speaks of a different kind of power. Certainly Foucault's excoriation of the totalizing and absolutist impulse of all retrospective 'total' histories and absolutist grand sociology (Foucault, 1972) confounds attempts to synthesize his work with that of Marx. For discussions of the implications of Foucault's analyses for Marxian theory see Poster (1984) and Smart (1983).

16. See Lowman and Menzies (1987, 115) for a graphic depiction of the decarceration of asylums in Canada and the US.

17. Young (1986, 9) terms this 'administrative criminology'.

18. The 'realist' interpretation of crime statistics should not be confused with the left realist interpretation as used in this context - the left realist interpretation acknowledges that both control factors and offender behaviour influence fluctuations in crime statistics. The relative contribution of each set of factors must remain an empirical question.

19. Ditton also contradicts himself. His juristic conception maintains that court decisions are the effective 'cause' of crime. Yet his definition of 'control' is clearly not

limited to crime since his empirical analysis deals with petty pilferage in a bakery. In this arena of private justice (cf. Henry, 1983, 1987) Ditton tells us that control rarely leads to official criminal proceedings (i.e. petty pilferage is rarely called a 'crime'). Official crime statistics therefore tell us nothing about the exercise of control in this context!

20. This is not to argue that 'fantasy rises' do not occur; for an example of one such effect see Schneider (1976) on the influence of neighbourhood watch progammes on burglary reporting rates.

21. For an alternative (but in some ways complementary) discussion of the politics of reform see Cohen's discussion of 'moral pragmatism' (1985, Ch. 7), and his critical, but supportive, assessment of left realism (1987).

22. Elsewhere Young (1986, 23) notes that his call to recognize that crime really is a problem 'is not to deny the impact of crimes of the powerful or indeed of the social problems created by capitalism which are perfectly legal. Rather, left realism notes that the working class is a victim of crime from all directions'. He also notes that left realism should not be construed as a simple empiricism (1986, 24).

23. Not that I have ever recommended doing so despite Harries's (1986, 578) misrepresentation to the contrary.

24. Respondents to this survey 'assigned an index of 2.1 to activity described as "a woman engages in prostitution", compared to 43.2 for "robbing a victim at gunpoint. The victim struggles and is shot to death"' (Harries 1986, 578).

25. In fact, in most cultures, but particularly in western cultures, it is almost impossible to understand the social control of women in isolation from the social control of their sexuality (cf. Bleier 1984, 180-90).

26. For a review of American research on street prostitution see Weisberg (1985) and of Canadian research see Lowman (1987a).

27. Although it should be noted that Revenue Canada rarely uses criminal proceedings against tax evaders (in contrast to the US).

28. For an introduction to crimes of the powerful, see Michalowski (1985, 314-401).

29. I do not take 'women's studies' to be a separate discipline, despite the marginalization of feminist perspectives into 'women's studies' departments in some

universities. Rather, feminist perspectives (at this particular historical juncture) are a necessary component of any critical theory (see also Davis and Faith 1987).

References

Biderman, A.D. and Reiss, A.J. (1967) 'On exploring the dark figure of crime', Annals, American Academy of Political and Social Science, 374, 1-15

Bland, L. (1984) 'The case of the Yorkshire Ripper: mad, bad, beast or male', in P. Scraton and P. Gordon (eds) Causes for Concern: British Criminal Justice on Trial, Harmondsworth: Penguin, 184-209

Bleier, R. (1984) Science and Gender: A Critique of Biology and its Theories on Women, New York: Pergamon Press

Bottomley, A.K. and Coleman, C.A. (1981) Understanding Crime Rates: Police and Public Roles in the Production of Official Statistics, Farnborough, Hants: Saxon House

Brake, M. (1980) The Sociology of Youth Culture and Youth Subculture, London: Routledge & Kegan Paul

Brooks, N. and Doob, A. (1986) 'Controlling income tax evasion', paper presented at the meeting of the Canadian Institute of Advanced Research's program on Sanctions and Rewards, University of Toronto Law School, December 16

Chunn, D.E. and Gavigan, S.A.M. (1986) 'Social control: analytical tool or analytical quagmire?', paper presented at the Annual Meeting of the Canadian Association of Anthropology and Sociology, 4-7 June, Winnipeg, Manitoba

Cohen, S. (1985) Visions of Social Control, Cambridge, Polity Press

Cohen, S. (1987) 'Taking decentralization seriously: values, visions and policies', in J. Lowman, R.J. Menzies, and T.S. Palys, Transcarceration: Essays in the Sociology of Social Control, Aldershot: Gower, 358-79

Cohen, S. and Scull, A. (1983) 'Social control in history and sociology', in S. Cohen and A. Scull (eds), Social Control and the State, Oxford: Martin Robertson, 1-14

Cousins, M. and Hussain, A. (1984) Michel Foucault, London: Macmillan

Davis, N.J. and Faith, K. (1987) 'Women and the state: changing models of social control', in J. Lowman, R.J. Menzies, and T.S. Palys (eds) Transcarceration: Essays in the Sociology of Social Control, Aldershot: Gower, 170-87

Ditton, J. (1979) Controlology: Beyond the New Criminology, London: Macmillan

Ellis, D. (1987) The Wrong Stuff: An Introduction to the Sociological Study of Deviance, Don Mills, Ontario: Collier Macmillan

Ericson, R. (1982) Reproducing Order: A Study of Police Patrol Work, Toronto: University of Toronto Press

Foucault, M. (1972) The Archaeology of Knowledge and the Discourse on Language, New York: Pantheon Books

Foucault, M. (1977) Discipline and Punish: The Birth of the Prison, London: Allen Lane

Garland, D. (1985) Punishment and Welfare, Aldershot: Gower

Gordon, C. (1980) Power/Knowledge: Selected Interviews and Other Writings 1972-1977, Michel Foucault, New York: Pantheon Books

Gregory, D. (1978) Ideology, Science and Human Geography, London: Hutchinson

Harries, K.D. (1980) Crime and Environment, Springfield, Illinois: Charles C. Thomas

Harries, K.D. (1986) 'Commentary on the geography of social control', Annals of the Association of American Geographers, 76, 4, 577-9

Henry, S. (1983) Private Justice: Towards Integrated Theorizing in the Sociology of Law, London: Routledge & Kegan Paul

Henry, S. (1987) 'The construction and deconstruction of social control: thoughts on the discursive production of state law and private justice', in J. Lowman, R.J. Menzies, and T.S. Palys, Transcarceration: Essays in the Sociology of Social Control, Aldershot: Gower, 89-108

Herbert, D.T. (1976) 'The study of delinquency areas: a geographical approach', Transactions of the Institute of British Geographers, 1, 472-92

Herbert, D.T. (1977) 'Urban crime: a geographical perspective', in D.T. Herbert and D.M. Smith (eds) Social Problems and the City: Geographical Perspectives, Oxford: Oxford University Press, 117-38

Hirschi, T. (1969) Causes of Delinquency, University of California Press, Berkeley, California

Hobsbawm, E. (1968) 'Cities and insurrections', Architectural Design, 38, 579-88

Ignatieff, M. (1978) A Just Measure of Pain: The

Penitentiary in the Industrial Revolution, London: Macmillan
Janowitz, M. (1975) 'Sociological theory and social control', American Journal of Sociology, 81, 1, 82-108
Keat, R. and Urry, J. (1975) Social Theory as Science, London: Routledge & Kegan Paul
Kinsey R., Lea, J., and Young, J. (1986) Losing the Fight Against Crime, Oxford: Blackwell
Kitsuse, J.I. and Cicourel, A.V. (1963) 'A note on official statistics', Social Problems, 11, 4, 131-9
Kress, J.M. (1980) 'The spatial ecology of criminal law', in D.E. Georges-Abeyie and K.D. Harries (eds) Crime: a Spatial Perspective: 58-71, New York: Columbia University Press
Lea, J. and Young, J. (1984) What is to be Done About Law and Order: Crisis in the Eighties, Harmondsworth: Penguin in association with the Socialist Society
Lowman, J. (1982) 'Crime, criminal justice policy, and the urban environment', in D.T. Herbert and R.J. Johnston (eds) Geography and the Urban Environment: Progress in Research and Applications, vol. 5, 307-42, Chichester: John Wiley and Sons
Lowman, J. (1983) 'Geography, crime and social control', unpublished Ph.D. dissertation, Department of Geography, University of British Columbia
Lowman, J. (1984) Vancouver Field Study of Prostitution, two volumes, Working Papers on Pornography and Prostitution, Report No. 8, Department of Justice, Ottawa
Lowman, J. (1986a) 'Conceptual issues in the geography of crime: toward a geography of social control', Annals of the Association of American Geographers, 76, 1, 81-94
Lowman, J. (1986b) 'Reply to Keith Harries, "Commentary on the geography of social control"', Annals of the Association of American Geographers, 76, 4, 579-81
Lowman, J. (1986c) 'Prostitution in Vancouver: some notes on the genesis of a social problem', Canadian Journal of Criminology, 28, 1, 1-16
Lowman, J. (1987a) 'Taking young prostitutes seriously', Canadian Review of Sociology and Anthropology, 24, 1, 99-116
Lowman, J. and Menzies, R.J. (1987) 'Out of the fiscal shadow: carceral trends in Canada and the United States, Crime and Social Justice, 26, 95-115
Lynch, M.J. and Groves, W.B. (1986) A Primer in Radical

Criminology, Albany, New York: Harrow and Heston
McDonald, L. (1979) The Sociology of Law and Order, Toronto: Methuen
MacIver, R.M. and Page, C. (1949) Society, London: Macmillan
Mayer, J.A. (1983) 'Notes towards a working definition of social control in historical analysis', in S. Cohen and A. Scull, Social Control and the State, Oxford: Martin Robertson, 17-38
Meier, R.F. (1982) 'Perspectives on the concept of social control', Annual Review of Sociology, 8, 35-55
Melossi, D. and Pavarini, M. (1981) The Prison and the Factory: Origins of the Penitentiary System, London: Macmillan
Melossi, D. (1987) 'The law and the state as practical rhetorics of motive: the case of "decarceration"', in J. Lowman, R.J. Menzies, and T.S. Palys (eds) Transcarceration: Essays in the Sociology of Social Control, Aldershot, Gower, 27-42
Michalowski, R.J. (1985) Order, Law and Crime: An Introduction to Criminology, New York: Random House
Mills, C.W. (1943) 'The professional ideology of social pathologists, American Journal of Sociology, 49, 165-80
Monahan, J., Novaco, R., and Geis, G. (1979) 'Corporate violence: research strategies for community psychology', in D. Adelson and T. Sorbin (eds) Challenges for the Criminal Justice System, New York: Human Sciences Press
Park, R.E. (1921) Introduction to the Science of Sociology, Chicago: University of Chicago Press
Parsons, T. (1951) The Social System, New York: Free Press
Poster, M. (1984) Foucault, Marxism and History: Mode of Production versus Mode of Information, Cambridge, Polity Press
Pound, R. (1942) Social Control Through Law, Connecticut: Yale University Press
Radcliffe-Brown, A.R. (1933) 'Social sanctions', Encyclopaedia of the Social Sciences, 13, 531-4, New York: Macmillan
Rafter, N.H. (1986) 'Left out by the left: crime and crime control', Socialist Review 16, 5, 7-23
Reckless, W. (1961) 'A new theory of delinquency and crime', Federal Probation, 25, 42-6
Reiss, A. (1951) 'Delinquency as a failure of personal and social control', American Sociological Review, 16, 2,

196-206

Ross, E.A. (1901) Social Control: A Survey of the Foundations of Order, New York: Macmillan

Rothman, D.J. (1971) The Discovery of the Asylum: Social Order and Disorder in the New Republic, Boston: Little Brown

Rusche, G. and Kirchheimer, O. (1939) Punishment and Social Structure, New York: Russell and Russell

Schneider, A.L. (1976) 'Victimization surveys and criminal justice system evaluation', in W.G. Skogan (ed.) Sample Surveys of the Victims of Crime, Cambridge, Mass: Ballinger

Schwendinger, H. and Schwendinger, J. (1974) Sociologists of the Chair, New York: Basic Books

Scull, A. (1977) Decarceration: Community Treatment and the Deviant: A Radical View, Englewood Cliffs, N.J.: Prentice Hall

Silbert, M.H. and Pines, A.M. (1981) 'Occupational hazards of street prostitutes', Criminal Justice and Behavior 4, 4, 395-9

Smart, B. (1983) Foucault, Marxism and Critique, London: Routledge & Kegan Paul

Smith, S.J. (1986) Crime, Space and Society, Cambridge: Cambridge University Press

South, N. (1987) 'The security and surveillance of the environment', in J. Lowman, R.J. Menzies, and T.S. Palys (eds) Transcarceration: Essays in the Sociology of Social Control, Aldershot: Gower, 139-52

Stang Dahl, T.S. and Snare, A. (1978) 'The coercion of privacy: a feminist perspective', in C. Smart and B. Smart (eds) Women, Sexuality and Social Control, London: Routledge & Kegan Paul

Stinchcombe, A. (1966) 'Institutions of privacy in the determination of police administrative practice', American Journal of Sociology, 69, 150-60

Sutherland, E. (1939) Principles of Criminology, 3rd edn, Philadelphia: Lippincott

Sutherland, E. (1945) 'Is "white collar crime" crime?', American Sociological Review 10, 132-9

Taylor, I.R. (1981) Law and Order: Arguments for Socialism, London: Macmillan

Tonnies, F. (1887) Gemeinschaft und Gesellschaft, Leipzig: Reisland

Turk, A. (1969) Criminality and Legal Order, Chicago: Rand McNally

Vold, G. (1958) Theoretical Criminology, New York: Oxford University Press

Weisberg, D.K. (1985) Children of the Night: A Study of Adolescent Prostitution, Lexington: D.C. Heath

Whyte, W.F. (1943) 'Social organization in the slums', American Sociological Review 8, 34-9

Young, J. (1979) 'Left idealism, reformism and beyond: from new criminology to Marxism', in National Deviancy Conference/Conference of Socialist Economists, Capitalism and the Rule of Law, London: Hutchinson, 11-28

Young, J. (1986) 'The failure of criminology: the need for a radical realism', in R. Matthews and J. Young Confronting Crime, London: Sage, 4-30

Chapter Twelve

POLICING AND THE CRIMINAL AREA

R.I. Mawby

Introduction

The geography of crime - the location of offenders and offences within physical and social space - is ultimately dependent upon the validity of measures of crime, and especially upon police statistics. The role of the police in the creation of official statistics is thus a key issue which must be confronted by any geographer who wishes to explain the spatial distribution of crime.

That said, academic interest in the policing of different areas has been varied. It is, for example, no more than fifteen years since pervading ideologies within sociology implicitly assumed that the concept of high-crime areas was problematic, and that variations between areas in levels of recorded crime were almost entirely accounted for by differential policing. Today, with fear of crime and risk more clearly identified through both national and local victim surveys, the pervasiveness and immediacy of crime for the disadvantaged in the inner city appears to be accepted almost uncritically.

In this paper, I shall therefore attempt, albeit briefly, to describe what I identify as five phases in the relationship between the geography of crime and policing practices. The first, the pre-academic phase, focuses on writings in the nineteenth century in which police practices were seen as peripheral to the distribution of crime. That is, area variations based on official statistics were considered unproblematic. This view is repeated in the second phase, the ecology phase, in which the problems of using police data are acknowledged but generally not confronted. In contrast, the third phase, in which crime data are seen as an artefact of gatekeeping, identifies those who saw police discretion as undermining the validity of crime statistics. The fourth phase, centring on my own work, is defined as a corrective on police influence, where the limitations to police discretion are considered. Finally, in the context of current developments, a fifth phase is suggested, where the emphasis is on the quality of policing in the community.

The bulk of this paper is thus divided into five sections, focusing on these distinct phases. However, given the emphasis within the policing research, more attention will be devoted to the third and fourth phases.

The pre-academic phase

Until the 1960s, discussions of area crime tended either to accept definitions of high-crime areas as unproblematic, or to concede that policing practices were of some influence but assume them to be minimal.

Writers, practitioners, and politicians in Victorian Britain, for example, had few doubts that offenders tended to congregate in particular areas, the 'rookeries', and that consequently these locations were virtually no-go areas for the respectable classes. Those whose work - or vocation - took them into such areas wrote vividly of their experiences. Dickens, for example, recounted with pride his adventures on patrol with Inspector Field. A London city missionary, writing slightly later, was somewhat less enthusiastic about his mission to the heathens of St Giles:

> So criminal and reckless is the population that the police always come here in couples, and at some seasons in bands of twelve. Even the city missionary's life is often endangered by the violence of the Irish, who have a way of pressing you against a wall, or of throwing you down and dancing on you, which is anything but comfortable.
>
> (London City Mission Magazine 1870, 57)

In this context, writers had little hesitation in identifying crime as spatially located, and developed a number of theories to account for such concentrations of juvenile or adult crime: density, deprivation, physical conditions, lack of education or religious influence, poverty, or drink, for example.

In such accounts, the role of the police was scarcely mentioned, although there was some appreciation that historically the lack of police presence in certain areas had contributed to these areas becoming magnets to those wanted by the police:

> It seems to have been the heritage of Westminster, by the force of custom, after the institution of the Sanctuary and the formation of its 'Thieves' Lane', to become the shelter and resort of lawless characters, who find a fitting home in the dirty, narrow, uncleansed streets - its miserable, undrained, dilapidated courts and alleys, reproduced and rebuilt time after time with the determinate purpose of receiving only the degraded and outcast of the population.
>
> (Beames 1852, 122)

Thus, rather than an excessive policing presence leading to exaggerated crime statistics, a lack of police presence was seen as an encouragement to offenders seeking a safe haven. Overall, though, policing practices were seen as peripheral to the distribution of crime. To the Victorian middle-class reformers and philanthropists, spatial variations in street safety were self-evident.

The ecological phase

Chicago in the 1920s showed many of the problems prevalent in Victorian London. It was expanding with the migration of rural and foreign born groups, distinctions between different parts of the city were emerging, and vice and crime were identified as particular problems. In this context, the Chicago school of Park, Burgess, and their colleagues presented a model which attempted to describe and explain urban diversity. Specifically, Shaw and McKay mapped crime rates in different zones of the city and related variations to the key concepts of the Chicago model, with the zone of transition identified as the centre of 'demoralization, of promiscuity, and vice' (Burgess 1925, 59).

Shaw and McKay (1969) in fact concentrated on three sets of official statistics for continuous periods from 1900 onwards: juveniles taken before the Juvenile Court on delinquency petitions, delinquents committed to correctional institutions, and alleged delinquents dealt with by police and probation with or without a court appearance.

The extent to which these statistics are subject to gatekeeper discretion is scarcely mentioned in the key text. It is however, covered by implication in much of the research emanating from this statistical base. Thus, a number of studies, based on observation of street life in specific areas of the city, provide excellent ethnographic accounts of the social environment within which delinquency takes place, and where much of the crime described goes unreported or undetected (Thrasher 1927; Whyte 1964; Spergel 1964). This said, however, the charge that area variations are to a large extent dependent upon policing practices is most notably addressed in writings which post-date, and indeed are a response to, the critique of labelling theorists and symbolic interactionists (Cohen 1965).

If there is little awareness in the writings of the Chicago school of the problematic qualities of official statistics, much the same can be said of the plethora of British studies which emerged in the post-war period. For example, Mannheim's (1948) Cambridge research and Jones's (1958) comparison of council estates in Leicester, are dependent upon official statistics, and - even more contentious - Morris's (1957) Croydon research, often considered the classic British area study, is dependent upon probation statistics. Similarly, research by geographers, emerging in the 1970s, while not exclusively based on official statistics, tends to avoid consideration of police influence on crime patterns (Davidson 1981; Evans 1980; Herbert 1977).

Crime statistics as an artefact of the policing process

A radical shift in emphasis was heralded by the emergence of symbolic interactionism and labelling theories as dominant within the sociology of deviance of the mid-1960s. However, it would be wrong to assume that these critiques focused on area studies. On the contrary, the assault was fired at conventional criminology, with its atheoretical and reformist tradition, and indeed a subsequent 'body-count' suggests that compared with others the Chicago school, with its ethnographic limbs, suffered a relatively minor mauling.

The critique of official statistics drew much from Douglas's (1967) work on suicide and Garfinkel's (1967) critique of medical records. Applying these arguments to crime statistics, Kitsuse and Cicourel argued that official statistics tell us more about the law-enforcement processes than they do about the distribution of activity defined a priori as illegal:

> Thus, the questions to be asked are not about the appropriateness of the statistics, but about the definitions incorporated in the categories applied by the personnel of the rate-producing social system to identify, classify and record behaviour as deviant ... Rates can be viewed as indices of organizational processes rather than as indices of the incidence of certain forms of behaviour.
>
> (Kitsuse and Cicourel 1963, 131)

Similarly, in Britain:

> The practical application of law depends very largely on bureaucratically organized agencies charged with its enforcement, and it is from the information provided by those agencies that criminal statistics are produced. The scope of the data will therefore be bounded by the scope of a particular organization's activities, and the quality of the data, by the efficiency of that organization. The information will be above all else a record of the day to day activities of the agency concerned.
>
> (Wiles 1975, 212)

To a large extent, such critiques accredited the police with a degree of discretionary power which in retrospect was realized to be unrealistic. Nevertheless, bolstered by early sociological studies of the police, especially of policing 'victimless crimes' (drug offences for example) (Skolnick, 1966), sociologists unequivocally identified the police as the most powerful gatekeepers, empowered with crucial decisions over what to investigate, whom to caution, and decisively, whom to arrest. This is perhaps most graphically illustrated by Werthman and Piliavin (1967, 56):

> From the front seat of a moving patrol car, street life in a typical Negro ghetto is perceived as an uninterrupted sequence of suspicious scenes. Every well-dressed man or woman standing aimlessly on the

> street during hours when most people are at work is carefully scrutinized for signs of an illegal source of income; every boy wearing boots, black pants, long hair and club jacket is viewed as potentially responsible for some item on the list of muggings, broken windows, and petty thefts that still remain to be cleared; and every hostile glance directed at the passing patrolman is read as a sign of possible guilt.

Despite this, however, the spatial impact of policing received rather less attention than did the effects of police strategies on the class and race of 'known' offenders. For example:

> (B)oth the decision made in the field - whether or not to bring the boy in - and the decision made in the station - which disposition to invoke - were based largely on cues which emerged from the interaction between the officer and the youth, cues from which the officer inferred the youth's character. Those cues included the youth's group affiliations, age, race, grooming, dress, and demeanor. Older juveniles, members of known delinquent groups, Negroes, youths with well-oiled hair, blackjackets, and soiled denims or jeans (the presumed uniform of the 'tough' boys), and boys who in their interactions with officers did not manifest what were considered to be appropriate signs of respect, tended to receive the more severe dispositions.
>
> (Piliavin and Briar 1964, 210)

There are of course exceptions. Cicourel (1976) for example, is sometimes cited for showing the invalidity of area police statistics. However, Cicourel's study of juvenile justice is more limited in this respect, illustrating merely that delinquents who are prosecuted may vary in a number of social characteristics in different force areas or over time if the police adopt different policies or practices. Clearly this is of significance to macro-area studies, like that of Harries (1974) but tells us nothing of area differences between urban neighbourhoods within a force area.

A more localized perspective is indeed to be found in a British study of the time, by Armstrong and Wilson (1973) in the Easterhouse district of Glasgow. Focusing on the processes by which Easterhouse attained its reputation as Glasgow's major crime locale, the authors describe the roles of the press and the police in the amplification process. In the latter case, they suggest that the introduction of proactive policing, with squads (the 'Untouchables') brought into the area from outside, led to the over-recording of crime in the area, and the increased likelihood of youths in Easterhouse - compared with elsewhere - being arrested for ambiguous street incidents. For example:

> The appearance of the Untouchables ... coincided with a widening of definitions of 'delinquent behaviour', thereby placing more youths at risk in terms of being officially labelled.
>
> (Armstrong and Wilson 1973, 80)

Nevertheless, Armstrong and Wilson do not deny that Easterhouse had a high crime rate, only that it was not worse than many other areas in Glasgow which received less publicity, less police activity and consequently relatively lower recorded crime rates.

In fact, a more promising line of enquiry is suggested by Stinchcombe (1963) and Duster (1970) in considering the relationship between policing and public/private space. They rightly observe that police influence is limited where illegality occurs in private, especially non-visible space, but is more evident in public space. Thus offences committed in public, or suspects who spend more time in public space, are more vulnerable. In particular, the tramp:

> Few of us ever see a policeman in those places where we spend most of our time; a 'tramp' sees one wherever he goes, and the policeman has the discretionary power to 'run him in'.
>
> (Stinchcombe 1963, 152)

Now this raises a number of issues of relevance to the sociologist or geographer, about the relationships between socio-economic status, private space, and crime. Certainly much conventional crime, conventionally committed by the lower classes, either occurs in public or is made public by an interested party (usually the victim), whilst those white-collar offences traditionally associated with the middle classes are less visible, the more so in the high-tech age.

However, labelling theorists tended to view such class

differences on a socio-political level (middle-class crimes were not defined as 'real' crimes and thus not investigated) rather than on the level of practical policing.

In terms of the relationship between area crime rates and policing, in fact, the most comprehensive research emerged not from a sociologist of deviance, but from a political scientist of the right, J.Q. Wilson. Indeed, Wilson (1968) remains the most detailed study of the impact of different policing styles on area crime.

Essentially, Wilson distinguished three different policing styles, which might vary between departments, stations, or over time:

(i) The legalistic style, with an emphasis on minimum discretion: those who were identified as law-breakers were routinely prosecuted.

(ii) The watchman style, with an emphasis on order-maintenance: illegal action did not necessarily result in arrest, but rather police action depended on how best order might be maintained or restored.

(iii) The service style, with an emphasis on the police role as a helping one: police action was thus taken in the context of the wider interests of the community.

The implication of this for the relationship between recorded crime rates and policing practices is considerable. In line with labelling theorists and interactionists, Wilson associated 'service style' policing with middle-class suburban areas, the implication being both that the service provided by the police in such areas will be distinctive and that as a result recorded offending by locals will be comparatively low. However, and echoing Bittner (1967), Wilson argues that under-recording will also be relatively common in inner-city, transient areas, where the 'watchman style' of policing priorities emphasizes 'order-maintenance' rather than 'law-enforcement' practices. Thus, the simple linear relationship between socio-economic status/reputation of an area and the extent to which the police use their discretion is rejected for a more complex set of relationships.

Nevertheless, Wilson's results gave interactionists further substance for the assumption that official crime statistics on an area basis should be viewed with scepticism, if not total disbelief. It was in this atmosphere that the

Sheffield research on urban social structure of crime emerged in the late 1960s and early 1970s. The first stage, described by Baldwin and Bottoms (1976) while based on official statistics, included an appreciation of the questionable validity of such data; the second stage subsequently focused on the influence of police practices on area recorded crime rates.

The creation of area crime rates: a corrective

The second stage of the Sheffield project focused on the processes whereby crime came to be reported to and detected by the police, and specifically on differences between residential areas. Nine areas were consequently chosen, varying in terms of their crime rates, housing design (houses or flats), and tenure type (owner-occupied, privately rented, or council). There was a positive relationship between offence rates (based on the area in which the offence occurred) and offender rates (based on the area in which the offender lived), both for these nine areas and overall for Sheffield, but it must be stressed that this did not mean that most offences were committed by locals or that locals committed the majority of their offences on their 'home patches' (Mawby 1979).

The most important findings from this stage of the research indeed suggested tht the impact of policing strategies on area crime rates was minimal. Following Reiss (1971) and in direct contrast to the interactionist tradition, it was evident that most crime was reported to the police, not discovered directly by them. Moreover, while there were variations by offence type and offender characteristics (notably age), the extent of proactive policing varied only marginally according to the crime rates of the areas (as officially defined). Where there were area variations, these were between residential and commercial/industrial districts, not within the residential sector. Furthermore, other records on crime and deviance - collected by the Post Office (Mawby, 1977, 1978a) or Housing Department (Xanthos, 1981) - illustrated the fact that areas with high crime rates according to police records tended to have high rates on quite independently collected data.

A number of reasons appeared to account for these findings. First, clearly, the extent to which crime data were dependent upon public reporting limited police influence.

Second, it seemed that for such relatively small areas individual police officers had rather vague views of the crime status of their 'patches' compared to others especially others of similar socio-economic or tenure status (Mawby 1979). Third, since police discretion generally operated where the offence occurred rather than where the offender lived, the influence of policing strategies was more apparent for offence patterns than offender patterns.

The third stage of the Sheffield project incorporated participant observation (Xanthos 1981) a self-report survey of offending by and victimization of juveniles from three of the areas (Mawby 1976, 1978b) and a victim survey (Mawby 1986a, Bottoms et al. 1987). Although the picture was not totally consistent - most notably in the two areas of high-rise flats - the overall pattern was broadly maintained. Namely, areas with contrasting crime rates according to official statistics were similarly distinctive according to self-report data, victim survey measures, and evidence from participant observation. At the time, I therefore concluded:

> (I)t is evident that the recorded information shows no indication of area differences being radically altered due to the different actions of the police (or indeed the public) in different areas.
>
> (Mawby 1979, 182)

Flying in the face of conventional wisdom, these conclusions were criticized by some criminologists at the time, (Ainsworth 1980) but the most concerted attack came from a geographer, John Lowman (Lowman 1982). Defending Cicourel and Matza, Lowman argued against my conclusions. For example, he pointed out that this stage of the Sheffield research was largely focused on working-class housing, and so avoided class-based comparison, and noted the lack of significant numbers from racial minority groups in Sheffield, compared with elsewhere. Two other criticisms raised by Lowman, that the research is ahistorical and based on indictable crimes are, however, misleading, being based on a simplistic reading of Policing the City and a lack of reference to other publications from the latter stages of the project (Mawby 1978c; Bottoms and Xanthos 1981). Since they are crucial issues in considering the influence of the police, it is perhaps worth spelling them out in some detail.

First, clearly the research was conducted at one point in time, and therefore refers to the effects of police practices on crime rates at that time. It does not consider the long-term effect of policing practices on crime rates, in terms of amplification processes, and indeed in this respect it is no different from the work of Cicourel or Matza. However, whilst the Sheffield research was recognized as time-specific (we could not discover when certain areas attained their problematic reputations in the inter-war period), all the evidence suggested that downward spirals in the publicly rented sector resulted from tenant selection/choices, where local authority housing policies affected the realistic choices open to tenants constrained by, for example, their power to wait for alternatives (Bottoms et al. 1987; Bottoms and Xanthos 1981). Thus, if any public agency were to be identified as having a significant influence on the creation of crime- (or problem-) prone environments, the housing department seemed a more influential agency than the police.

Secondly, Lowman somewhat mysteriously argues that the focus on indictable crime is misleading. In fact, Policing the City includes a wealth of additional police data, other than for motoring offences, most notably non-indictable offender data and incidents recorded by the police as non-crime (commonly neighbour or domestic disputes) (Mawby 1979). In some of these, certainly, police proactivity is more pronounced, although there is no evidence that where this is the case it distorts the overall picture. Moreover, it is quite clear that where incidents such as domestic disputes are highly dependent upon public reporting, but are not recorded as crimes by the police, a broadly similar spatial pattern emerges. That is, there is evidence that public reporting of such incidents produces marked area contrasts, and no evidence that the police record incidents from low-crime areas differently from those in high-crime areas.

In the context of research which focuses on police practices, the Sheffield research in fact complements Gill's (1977) study in a small area of Liverpool, called Luke Street. Although Gill's approach is centrally ethnographic, he does include a discussion of police arrest practices. Thus in an area with a high recorded offender rate, Gill notes that juveniles in the area were particularly vulnerable to arrest for minor street offences but that this tended to supplement other arrests. That is, rather than create a higher offender rate, police strategies resulted in the same offenders being arrested more frequently.

A recent study in the United States by Smith (1986) also

produces minimal evidence for suggesting that police practices create area variations in crime rates. Smith compares sixty neighbourhoods in three cities in terms of five measures of police behaviour. He concludes that proactive police investigations were most common in racially mixed areas and areas with high proportions of elderly residents but less likely in high-crime neighbourhoods. Moreover, both the likelihood of a report being written and an arrest being made varied by social status, not crime level, of an area, with arrests more likely in low-status neighbourhoods but written reports most common in high-status neighbourhoods. Indeed, controlling for other variables suggested that 'Victims in high-crime areas are less likely to have the incident reported by police' (Smith 1986, 333). Again then, while Smith's research does not directly address the influence of police behaviour on crime rates, the extent of variation in such behaviour between low and high crime-rate areas does not appear great.

However, the most telling rebuttal of the argument that police practices significantly influence area crime rates (or at least offence rates) comes, implicitly, from victimization surveys. These surveys, originating on a large scale in the United States in the 1960s, produce a record of crime irrespective of whether or not the police were involved. Yet, as has been shown on both a national and a local level, quite clearly crime is a more serious problem in areas of inner-city deprivation than in the suburbs or rural parts of the country. For example, on a national level, the ACORN classification of neighbourhood type is used by Hough and Mayhew (1985) to demonstrate that 'multi-racial areas' and the 'poorest council estates' both have high crime rates and include high proportions of residents who are concerned about crime.

Local studies make the same point even more forcefully. Thus as well as our own research in Sheffield, Kinsey notes marked area variations in Liverpool:

> There can be no doubt that both in terms of the quantity and impact of the crimes examined the poor suffer more than the wealthy. The problems appear critical for the 20 per cent of the Merseyside population living in the poorest council housing and especially in the District of Knowsley where 48 per cent of the population (six times the national average)

live in such areas.

(Kinsey 1984, 16)

A similar acceptance of a link between deprivation and victimization is to be found in Lea and Young (1984) and in the Islington crime survey, where the authors focus on the high crime risk facing the whole - but most especially certain parts - of Islington (Jones et al. 1986). Undoubtedly, as the crime problem has been introduced as an acceptable agenda by the left, so the assumption that the spatial location of offences is an artefact of differential policing, has been dismissed.

But where next? Certainly the assertion that crime rates do vary between urban locations shifts the focus away from seeing 'policing the criminal area' merely in terms of a 'police create crime rates' scenario. However, it should not be assumed that this is the end of the debate. Rather, those who have reaffirmed the vulnerability of the inner-city residents have gone on to focus on the quality of policing service. The final section thus briefly considers current interest in policing in different communities.

The quality of policing

One of the major concerns of nineteenth-century writers on crime and wider social policy was the changing nature of social control (Oben 1974). As cities grew, so did the extent of segregation. The masses thus lived their lives out of the paternal, controlling gaze of their betters. As one recent commentator has noted:

> The old methods of social control based on the model of the squire, the parson, face to face relations, deference and paternalism, found less and less reflection in the urban reality. Vast tracts of working-class housing were left to themselves, virtually bereft of any contact with authority except in the form of the policeman and the bailiff.
>
> (Stedman-Jones 1971)

Since then, the agents who imposed control on behalf of ruling elites, namely the police, have also become outsiders, residents of more desirable areas of the city who commute to work in crime-prone areas. In Plymouth for example, those areas where crimes most commonly occur and where known offenders reside are the more deprived neighbourhoods. The police, who are ultimately charged with control within such areas, live in the more prosperous suburbs (Mawby, 1986b). This is illustrated in Figures 12.1 and 12.2 where the distribution of reported household burglaries over a 13-week period contrasts with the private residences of a sample of 100 police officers.

What then is the nature of this 'control from without'? Whilst labelling theorists were convinced that the major problem was the over-concentration of the police in certain neighbourhoods (an issue which puzzled us in Sheffield where we could find no more than a handful of respondents - and these mainly students - who criticized the police for over-patrolling their area) more recent criticisms have focused on three aspects of policing. Specifically, the focus has been on the behaviour of individual police officers, the operational role of the police, and the nature of control over the police. The role of the police prior to and during inner-city riots, and the response of inquiries by those such as Lord Scarman (1981) to those riots, has set the agenda according to which such criticisms are made.

The behaviour and operational role of the police have been addressed on a number of levels, including criticism of the inadequacy of police protection in inner-city neighbourhoods (Jones et al. 1986) and criticisms of police involvement in service-type tasks rather than crime control (Kinsey 1984). However, where the police have been central in developing new initiatives based within the community, criticisms have centred upon these. For example, 'community policing' has been attacked as no more than an information-gathering exercise, whereby an outside agency (the police) 'con' the community into providing it with details which will ultimately be used against it and its interests (Gordon 1984).

A similar criticism has been levelled against neighbourhood watch. Imported from the United States, neighbourhood watch has developed relatively recently in a number of police areas, especially within the Metropolitan Police, with the aims of target hardening, property marking, and community involvement (Bennett 1987). As with community policing, however, it has been attacked as a means whereby the police increase their influence in the neighbourhood, but on their own terms, with control firmly located 'without' (Donnison et al. 1986). An additional

Figure 12.1 Distribution of burglaries in Plymouth, 1984–5

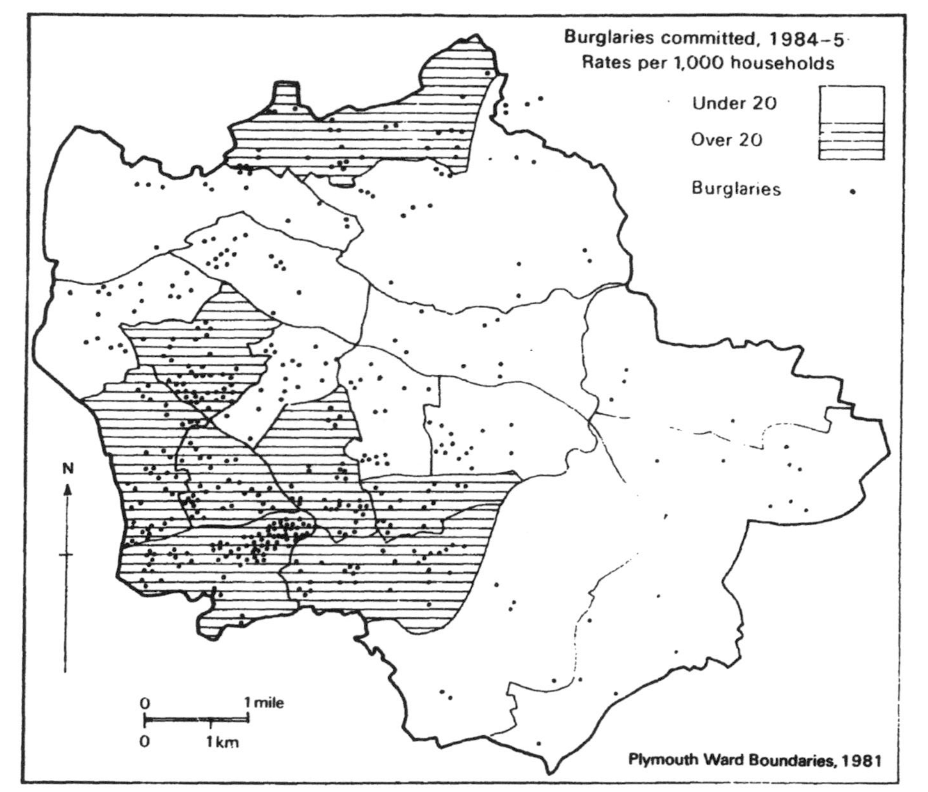

Figure 12.2 Residences of police officers in Plymouth, 1985

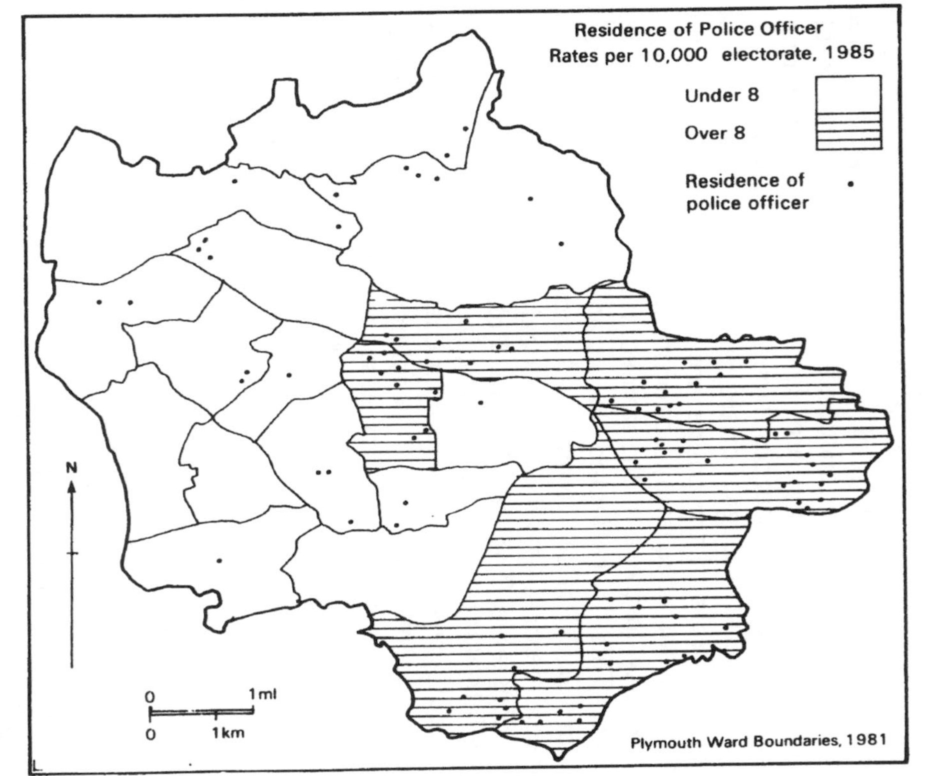

problem arises over the practicality of community involvement. If crime is in some way related to lack of community spirit, as the Chicago school argued, then this very lack of community cohesion would presumably undermine attempts to successfully run neighbourhood watch schemes. Not surprisingly, then, the evidence from both abroad (Lavrakas and Herz 1982) and nearer to home (Hourihan 1987; Bennion et al. 1985; Donnison et al. 1986) indicates considerable variations in individual and community commitment to neighbourhood watch. The result undermines the viability of schemes, and arguably means that schemes can most successfully operate in middle-class suburbs - where they are least needed - effecting a maldistribution of resources (Donnison et al. 1986).

In general though, criticisms of the behaviour and operational role of the police are built on concerns that the police are not adequately accountable to their publics, and particularly to those in the immediate neighbourhoods being policed. This is reflected in attacks on the complaints system (GLC Police Committee) and tripartite relationship between central government, local government, and the chief constable (Spencer 1985). Police responses, in terms of the introduction of lay visiting initiatives (Walklate 1987) and local consultative arrangements (Morgan 1987) are a partial acceptance of the problems posed. Nevertheless, what are of particular interest to geographers are the problems posed by such moves towards community involvement. For example, which members of the community are or ought to be involved, how are they selected, and as a result what mandate do they hold? Attempts to face the problems of policing in the community thus inevitably lead to a closer analysis of the nature of the so-called 'community' being policed.

There is, unfortunately, insufficient space here to consider these practical issues in more detail. What has, hopefully, been illustrated, however, is the shift in recent years away from criticisms of police saturation and towards monitoring the quality of and mandate for police services in the area.

Conclusion

This shift is in part political, with a noticeable willingness among those on the left to accept crime as a 'real' problem rather than myth (Lea and Young 1984; Downes 1983; Birley and Bright 1985) and an acceptance of the police as a necessary presence (Reiner 1985). For this reason, whilst some critiques of community policing (Gordon 1984) or surveillance techniques (Hain et al. 1980) may be interpreted as attacks on the police presence in working-class areas, such criticisms are more specifically about policing practices rather than policing in general.

To a large extent, this reflects not merely a changing awareness of the nature of the crime problems, but also a dramatic change of emphasis within research on policing. Thus, whereas early researchers saw police work as a theoretical or methodology issue, more recent studies have been policy-oriented. As a result, the debate is no longer about the impact of policing on official statistics and more about the effects of particular police practices on local problems. Geographers who read this article in future years may consequently see it as an historical oddity rather than a critique of a current issue.

References

Ainsworth, P.B. (1980) 'Review of policing the city', in International Journal of the Sociology of Law, 8, 463-6

Armstrong, G. and Wilson, M. (1973) 'City politics and deviancy amplification', in I. Taylor and L. Taylor (eds) Politics and Deviance Harmondsworth: Penguin

Baldwin, J. and Bottoms, A.E. (1976) The Urban Criminal, London: Tavistock

Beames, T. (1970) The Rookeries of London (1852), London: Cass

Bennett, T. (1987) 'Neighbourhood watch: principles and practice' in R. Mawby (ed.) Policing Britain, Plymouth: Plymouth Polytechnic

Bennion, C. (1985), 'Neighbourhood watch: the eyes and ears of urban policing', Occasional papers in Sociology and Social Policy, No. 6, Surrey: University of Surrey

Birley, D. and Bright, J. (1985) Crime in the Community, Labour Campaign for Criminal Justice, London

Bittner, E. (1967) 'The police on skid row', American Sociological Review, 32, 699-715

Bottoms, A.E. et al. (1987) 'A localized crime survey', British Journal of Criminology, 27, 125-54

Bottoms, A.E. and Xanthos, P. (1981) 'Housing policy and

crime in the British public sector' in P.J. Brantingham and P.L. Brantingham (eds) Environmental Criminology, London: Sage

Burgess, E. (1925) 'The growth of the city: an introduction to a research project' in R. Park et al. (eds) The City, Chicago: University of Chicago Press

Cicourel, A. (1976) The Social Organization of Juvenile Justice, London: Heinemann

Cloward, R. and Ohlin, L. (1960) Delinquency and Opportunity, Chicago: Free Press

Cohen, A. (1965) 'The Sociology of the deviant act: anomie theory and beyond', American Sociological Review, 30, 5-14

Davidson, R.N. (1981) Crime and Environment, London: Croom Helm

Dickens, C. 'On duty with Inspector Field', in Reprinted Pieces, 162-72, London: Odhams Press

Donnison, H. et al. (1986) Policing the People, London: Libertarian Research and Educational Trust

Downes, D. (1983) Law and Order: Theft of an Issue, London: Fabian Society

Douglas, J. (1967) The Social Meanings of Suicide, Princetown: Princetown University Press

Duster, T. (1970) The Legislation of Morality, New York: Free Press

Evans, D.J. (1980) Geographical Perspectives on Juvenile Delinquency, Aldershot: Gower

GLC Police Committee, (no date) The Police Act 1984, GLC Police Committee Discussion Paper No. 3, London

Garfinkel, H. (1967) Studies in Ethnomethodology, Harmondsworth: Penguin

Gill, O. (1977) Luke Street, London: Macmillan

Gordon, P. (1984) 'Community policing: towards the local police state', Critical Social Policy, 4, 39-58

Hain, P. et al. (1980) Policing the Police, vol. 2, London: Calder

Harries, K.D. (1974) The Geography of Crime and Justice, New York: McGraw-Hill

Herbert, D.T. (1977) 'Crime, delinquency and the urban environment', Progress in Human Geography, 1, 208-39

Hourihan, K. (1987) 'Local community involvement and participation in neighbourhood watch: a case study in Cork, Ireland', Urban Studies, 24, 129-36

Hough, M. and Mayhew, P. (1985) Taking Account of Crime: Key Findings from the Second British Crime Survey,

London: HMSO

Jones, H. (1958) 'Approaches to an ecological study', British Journal of Delinquency, 8, 277-93

Jones, T. et al. (1986) The Islington Crime Survey, Aldershot: Gower

Kinsey, R. (1984) Merseyside Crime Survey: First Report, Merseyside: Merseyside County Council

Kitsuse, J. and Cicourel, A. (1963) 'A note on the use of official statistics', Social Problems, 11, 131-9

Lavrakas, P.J. and Herz, E.J. (1982) 'Citizen participation in neighbourhood crime prevention', Criminology, 20, 479-98

Lea, J. and Young, J. (1984) What is to be Done about Law and Order?, Harmondsworth: Penguin

London City Mission Magazine (1870)

Lowman, J. (1982) 'Crime, criminal justice policy and the urban environment' in D.T. Herbert and R.J. Johnston (eds), Geography and the Urban Environment, vol. 5, Chichester: Wiley

Mannheim, H. (1948) Juvenile Delinquency in an English Middletown, London: Kegan Paul

Mawby, R.I. (1976) 'The victimization of juveniles', Journal of Research in Crime and Delinquency, 16, 98-113

Mawby, R.I. (1977) 'Kiosk vandalism', British Journal of Criminology, 17, 30-46

Mawby, R.I. (1978a) 'Policing by the Post Office', British Journal of Criminology, 18, 242-53

Mawby, R.I. (1978b) 'Crime and law-enforcement in residential areas of the City of Sheffield', PhD thesis, University of Sheffield

Mawby, R.I. (1978c) 'A note on domestic disputes', Howard Journal, 17, 160-8

Mawby, R.I. (1979) Policing the City, Aldershot: Gower

Mawby, R.I. (1986a) 'Contrasting measurements of crime rates: the use of official records and victim studies in seven residential areas' in K. Miyazawa and M. Ohya (eds) Victimology in Comparative Perspective, Tokyo: Seibundo

Mawby, R.I. (1986b) 'The geography of crime and the criminal justice system: gatekeepers as commuters' in D.T. Herbert et al. (eds) The Geography of Crime, Occasional Papers in Geography, Stoke: North Staffordshire Polytechnic

Morgan, R. (1987) 'Consultation and police accountability' in R.I. Mawby (ed.) Policing Britain, Plymouth: Plymouth

Polytechnic
Morris, T. (1957) The Criminal Area: A Study in Social Ecology, London: Routledge
Oben, T. (1974) 'Victorian London: specialization, segregation, and privacy', Victorian Studies, 17
Piliavin, I. and Briar, S. (1964) 'Police encounters with juveniles', American Journal of Sociology, 70, 206-14
Reiner, R. (1985) The Politics of the Police, Brighton: Wheatsheaf
Reiss, A.J. (1971) Police and the Public, Yale: Yale University Press
Scarman, Lord (1981) Report on an Enquiry by Lord Scarman: The Brixton Disorders, 10-12 April 1981, Cmnd 8427, London: HMSO
Shaw, C. and McKay, H. (1969) Juvenile Delinquency and Urban Areas, Chicago: University of Chicago Press
Skolnick, J. (1966) Justice without Trial, New York: Wiley
Smith, D.A. (1986) 'The neighbourhood context of police behaviour' in A.J. Reiss and M. Tonry (eds) Communities and Crime, Chicago: University of Chicago Press
Spencer, S. (1985) 'The eclipse of police authority' in B. Fine and R. Millar (eds) Policing the Miners' Strike, London: Lawrence and Wishart
Spergel, I. (1964) Racketville, Slumtown, Haulberg, Chicago: University of Chicago Press
Stedman-Jones, G. (1971) Outcast London, Oxford: Clarendon Press
Stinchcombe, A. (1963) 'Institutions of privacy in the determination of police administrative practices', American Journal of Sociology, 69, 150-60
Thrasher, E.M. (1927) The Gang, Chicago: University of Chicago Press
Walklate, S. (1987) 'Public Monitoring and Police Accountability', paper given to British Criminology Conference, Sheffield
Werthman, C. and Piliavin, I. (1967) 'Gang members and the police', in D. Bordua (ed.) The Police: Six Sociological Essays, New York: Wiley
Whyte, W. (1964) Street Corner Society, Chicago: University of Chicago Press
Wiles, P. (1975) 'Criminal statistics and sociological explanations of crime' in W. Carson and P. Wiles (eds) The Sociology of Crime and Delinquency in Britain, vol. 1, London: Robertson

Wilson, J.Q. (1968) Varieties of police behaviour, Harvard: Harvard University Press

Xanthos, P. (1981) 'Crime, the housing market and reputation', PhD thesis, Sheffield: University of Sheffield

Chapter Thirteen

CONTESTING CONSULTATION: THE POLITICAL GEOGRAPHY OF POLICE-COMMUNITY CONSULTATION IN LONDON

Nicholas R. Fyfe

This chapter describes the political response to a recent government policy requiring the police to consult local communities about the policing of their areas. Such a policy might seem unexceptional but in fact challenges many popular beliefs about British policing. It is often presumed, for example, that law enforcement is invariant over space and flows from the independent judgement of police officers; police-community consultation suggests, by contrast, the possibility of policing varying in response to the needs and wishes of local communities. Second, focusing on the political response to this policy might appear redundant given the belief that police and politics are mutually exclusive; the implementation of police-community consultation, however, demonstrates the way in which policing has become politicized within government. Both these issues suggest geographies of policing, one concerned with discretionary law enforcement, the other with policing as a political issue at different levels of government. The former has been considered in spatial studies of crime (Mawby 1979; Lowman 1982); the latter has only recently claimed the attention of geographers, notably Susan Smith, who has considered the role of public policy in crime prevention and the contribution of political geography to the debate about police accountability (Smith 1986a, 1986b).

In what follows I want to illuminate the substantive basis for a political geography of policing by describing the relations between police and politics within central, regional and local government. The discussion is focused around the implementation of the policy of police-community consultation in London, showing how the local form of consultation is the negotiated outcome of relations between different levels of government. The account is structured by considering in turn central, regional, and local government responses to consultation. I begin, however, by describing the origins of the policy in Lord Scarman's Report on the disorders in Brixton in 1981.

The Scarman Report and community-police consultation

During a weekend in April 1981 'the British people watched with horror and incredulity scenes of violence and disorder in their capital city, the like of which had not previously been seen in this century in Britain' (Scarman 1981, 1.2). These are the opening words of Lord Scarman's Report into the Brixton disorders. The Report combines an analysis of the causes of these disorders with a set of policy recommendations. Scarman concluded that the riots 'were essentially an outburst of anger and resentment by young black people against the police' (Scarman 1981, 3.110) and that a 'significant cause' of this hostility was a loss of confidence by the local community in the police, brought about in part by policing methods which failed to command the support of the local community (for example, stop and search, and saturation policing) and by the collapse of voluntary police-community liaison. This led him to the following conclusion and recommendation:

> If a rift is not to develop between the police and the public as a whole (not just members of ethnic minority communities), it is in my view essential that a means be devised of enabling the community to be heard not only in the development of policing policy but in the planning of many, although not all, operations against crime
>
> (Scarman 1981, 5.56)

The 'means' he recommended was the formation of local community-police consultative committees across England and Wales, composed of community representatives, local councillors, and police officers. This, he argued, would enhance the existing constitutional mechanism by which the police are made responsive to local communities, accountable to (partially) elected police authorities. Indeed, he believed this was urgently required in the Metropolitan Police District (MPD), covering Greater London, where its unique form of police governance (the police authority is the Home Secretary) means that the police are effectively insulated from any local democratic influence. Scarman acknowledged, however, that in London there was a demand for more radical reform of the accountability structure but refused to support any initiative which would see 'ultimate responsibility for the policing of the nation's capital transferred from a senior Minister responsible to (Parliament) and put in the hands of a local body' (Scarman 1981, 5.68). His only concession to London's peculiar constitutional position was to recommend that consultation take place at borough level rather than, as recommended for the rest of England and Wales, at police divisional or sub-divisional level. This territorial distinction is of particular political significance, however, because it gives borough councils a direct role in consultation.

Scarman's recommendation was eagerly taken up by central government and incorporated into the Police and Criminal Evidence Act (1984) (PACE). (1) Although it does not make consultative committees a statutory requirement, it does require police authorities to make appropriate arrangements for 'obtaining the views of the people' about the policing of their areas. After the 'scenes of violence and disorder' something was being done to reaffirm 'policing by consent', the hallmark of the British police tradition.

Police-community consultation in the Metropolitan Police District

Central government and consultation: the strategy from above

The government's commitment to implement Scarman's proposals on consultation was clearly signalled in London by the close personal involvement of the then Home Secretary, William Whitelaw, in establishing the Lambeth Community/Police Consultative Group. (2) This apparently unequivocal support, however, should not obscure the problems posed by the creation of a formal arena in which police officers would be expected to discuss in public local policing matters with elected politicians and community representatives, something without precedent in the MPD.

Indeed, there is a certain political irony that a

government so committed to defending the independent, professional judgement of police officers and to reducing the power and influence of councils in local affairs should implement a policy requiring the police to listen to the wishes of local communities. In part to counter this politically unpalatable interpretation of consultation central government drew up a set of guidelines for consultation. Outside London these guidelines have no statutory basis but in London the Metropolitan Police Comissioner, whose responsibility it is to establish consultation, has a statutory duty to consult them.

In effect then police-community consultation has been made a statutory requirement in the MPD. The guidelines set out the purpose and structure of consultation and indicate that themes informing government policy in other public services are to be introduced into policing. Consultative groups are to provide a mechanism to encourage a degree of self-help by informing communities of their responsibilities in crime prevention and should enhance police efficiency through consumerism by establishing some congruence between police policies and community wishes so that the public readily provide the information required to solve crimes (Morgan 1985). The guidelines also provide a strategy to restrict the role of local government in consultation by setting out: 'certain principles ... in relation to the membership and independence of the Group which must be adhered to if the spirit of consultation is to be implemented to the full' (Home Office 1985, 6).

These principles limited council representation to five members but placed no limit on community representation. This not only severely restricted the co-opted role of local authorities in negotiating locally 'appropriate' consultative arrangements but was seen by them to challenge the representational status of elected councillors. Indeed, these restrictions on local autonomy perhaps indicate that although central government was keen to present consultation simply as a mechanism for the fine-tuning of policing by consent, it was only too well aware of the political context within which consultation in London would quite literally take place: in the fragmented communities of the inner city where the gulf between police and sections of the community is at its greatest; and in the sprawling suburbs where many resented being treated as though they too had a 'problem' of police-community relations.

Figure 13.1 Police committees and police monitoring groups in London, 1987

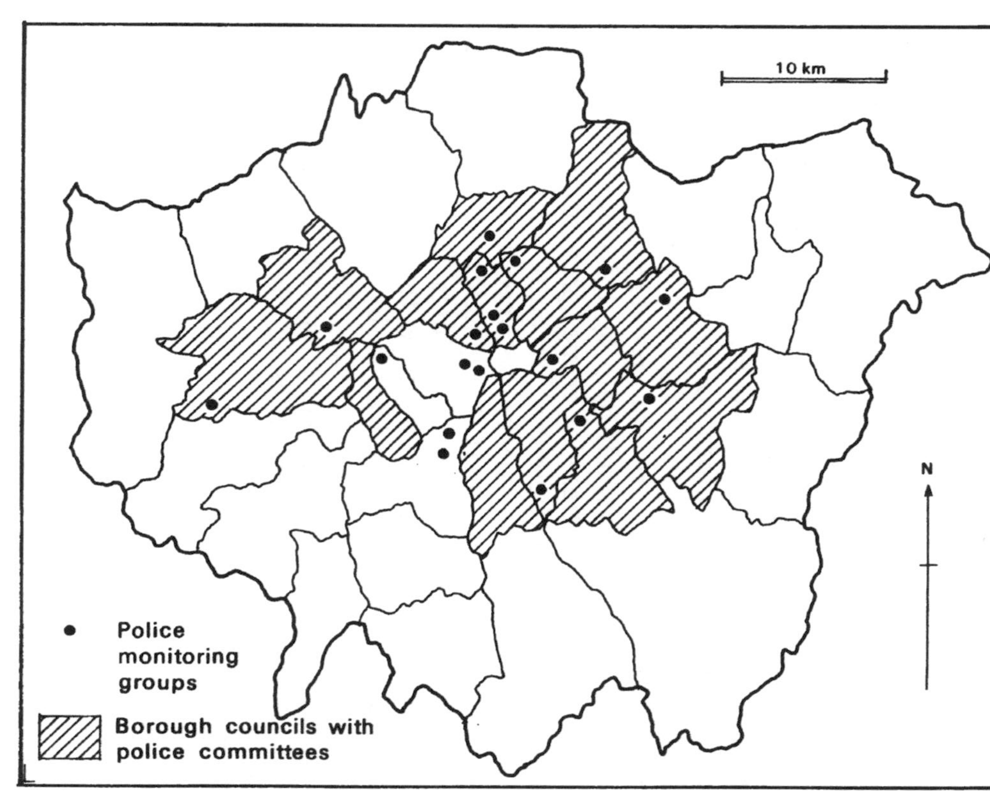

Regional government: the politicization of policing

In London policing has rarely been off the political agenda. Since the late nineteenth century there have been repeated demands for a locally elected police authority such as exist for the other forces of England and Wales (Bundred 1982). These demands have been rejected, the existing arrangement defended with the argument, reiterated by Scarman, that responsibility for policing the nation's capital cannot be given to a local body. Scarman's proposal for consultation was greeted angrily by the Labour-controlled Greater London Council (GLC), arguing that this 'fell far short of providing a satisfactory system for ensuring the accountability of the police to the community they serve' (GLC 1982, 115). The recently formed Police Committee of the GLC began to campaign vigorously for a democratically elected police authority, a campaign taken up at local level by sympathetic Labour borough councils forming their own police committees and by the GLC funding police monitoring groups. As Figure 13.1 illustrates the regional response to these intiatives has a political geography. It was the Labour-controlled inner-city councils which became most active in campaigning around policing issues; in much of the rest of London and particularly the Tory-controlled suburbs, policing was not a party-political issue. This regional political geography of policing is significant for in part it explains central government's strategy for implementing consultation. The guidelines had to be sufficiently strict to minimize the participation of those on inner London Labour councils whose political aspirations challenged the constitutional position of the Metropolitan Police; while being sufficiently flexible so as not to impose on those suburban and Conservative-controlled areas something which was seen as at best unnecessary, and at worst potentially subversive by attracting 'political trouble-makers'. Indeed, it is important to emphasize the resentment felt in many Tory areas at having to participate in an initiative which was prompted by riots in an inner-city Labour-controlled borough: 'They resent', in the words of a senior police officer 'the Lambeth tail wagging the consultative dog'.

It should not be presumed, however, that all those Labour councils setting up GLC-type police committees agreed with the GLC on the strategy for achieving the political goal of a democratically elected police authority for London. The consultative initiative illustrates a political divide between the GLC and some local Labour councils. The GLC's hostility towards consultation was partly based on the belief that it would undermine the campaign for full police accountability:

> the existence of these consultative committees can impede progress towards accountability in London by giving a false impression of control when in reality the police remain accountable to no one.
>
> (GLC 1983, 14)

At a local level many Labour councils who share the GLC's political aspirations have agreed to participate in consultation; and where they have adopted a hostile stance it has had less to do with police accountability and more to do with the way consultative groups would challenge the local autonomy of the council.

The local political response: contesting consultation

There has been a complex and diverse response to establishing community-police consultation across London. Although PACE gives statutory responsibilty to the Commissioner to set up groups, after taking account of Home Office guidance, he also has a duty under the Act to consult with each borough council as to locally appropriate arrangements. This has given these councils a pivotal role in determining the form of consultation in the boroughs. On the one hand, they have been negotiating with central government (via the Home Office) and the police over the constitution of the groups, while, on the other hand, they have been involved in selecting community groups for participation in consultation. The emerging pattern of consultation is shown in Figure 13.2 representing a 'snapshot' of the results of discussions over the local form and structure of consultation. Indeed, the pattern is very much a political geography, the negotiated outcome of the relations between local, regional, and central government. In accounting for the local response, however, no simple party-political explanation is possible: Conservative councils have been as hostile towards consultation as Labour councils, although clearly for different reasons. In Wandsworth, for example, local Tory councillors viewed consultation as subversive; in Lambeth, Labour councillors have boycotted

Figure 13.2 Police-community consultative groups in London, 1985

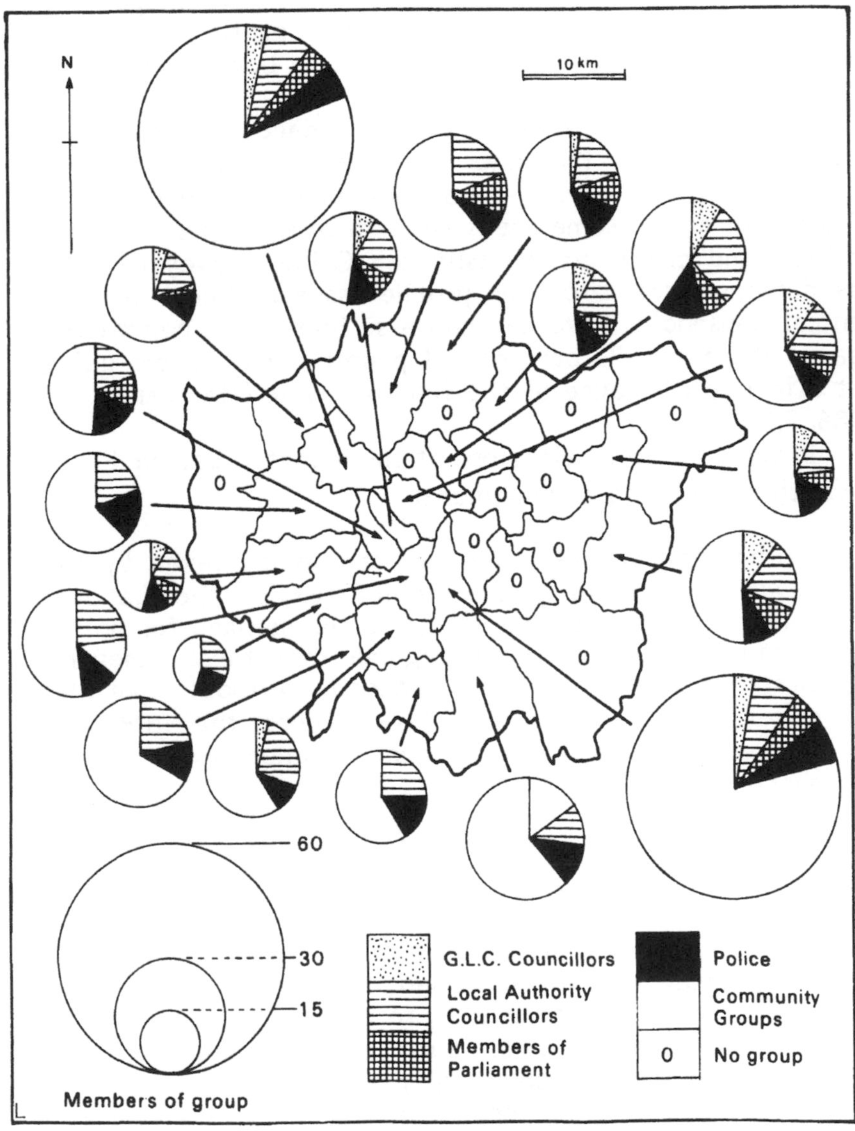

the group because it is perceived as unrepresentative and, echoing the GLC argument, obstructing moves towards a locally elected London police authority. In this section I want to indicate some of the issues which have shaped the local response by considering the implementation of police-community consultation in two contrasting London boroughs, Islington and Sutton.

Islington, contrary to the popular image of a gentrified 'yuppie' enclave in the inner city, is an area of severe deprivation. In the borough as a whole unemployment is over 20 per cent and in some areas rises to over 30 per cent. (3) Policing and crime, as the Islington Crime Survey of 1985 has demonstrated, are high-profile issues: almost 30 per cent of households have had had a serious crime committed against them, and 23 per cent of people always avoid going out after dark because of the fear of crime (Maclean et al. 1986, 7.2.).

In the 1982 local elections of the Islington Labour Party made this manifesto commitment:

> We particularly welcome the steps being taken by the newly established police committee of the Labour-controlled GLC to give effect to the demand for police accountability. But we recognize that these essential first steps will not succeed without parallel developments at borough level where the interface between the police and the community is at its sharpest.

This commitment, described in the local press as 'Labour pledges town hall grip on police', (4) was implemented when Labour gained power and established a police sub-committee. Consisting of ten councillors but with the power to co-opt local groups, its broad aims were to advocate democratic control of the police, to gather knowledge about policing in anticipation of such control, and to forge alliances with certain community groups around policing issues by, for example, considering racial harassment and crime on council estates.

This is the local political background to Islington council's response to central government's invitation to prepare joint proposals with the local police on consultation arrangements for the borough. Islington council's initiative was to argue for the acceptance of the Police sub-committee as the appropriate consultative arrangement. In

August 1982 the local police commander was informed by the council that this sub-committee was 'the means of communication between the local community (as represented by its elected borough Council) and the police'. (5) This claim that effective community representation can be achieved through elected councillors is clearly at odds with the conception of community representation set out in the guidelines based on the participation of non-elected community groups. Both conceptions are problematic: the former, by giving the prime role to those with an electoral mandate opens councillors up to a charge of elitism for it presumes they are able, either by themselves or through selecting groups for co-option, to represent all interests; the latter presumes a participatory political culture where all groups have an equal ability to organize and articulate their interests. Indeed, the Home Secretary's insistence on this second conception of community in the guidelines suggests it has an 'evangelistic quality' (Weatheritt 1986, 122). The value of concepts like 'community', as one commentator observes, lies in their appeal to the sacred aspects of tradition from which policing draws its legitimacy and 'To question the appropriateness of such concepts is thus to question the legitimacy to which many (policing) innovations make appeal and so may appear fundamentally subversive of the very purpose to which they are directed' (Weatheritt 1986, 123). It is the elitism contained in the Council's proposals with which the Home Office took issue. Police and community representatives would not be members as of right and so would not be meeting councillors on equal terms. Further, the Home Office wanted MPs and GLC councillors to be members of the Group. The council continued to negotiate, arguing that if they were to follow the guidelines they would simply appoint the same councillors to the sub-committee as the consultative group and that both MPs and GLC councillors had arenas in which to discuss national and regional policing issues. In April 1983 the council was made an offer by the Home Office allowing ten councillors but at least eleven community representatives and both MPs and GLC councillors. The response of the police sub-committee was triumphant: 'The principle the Council has wished to establish throughout that the Council represents the community and must be in a position to do so is given practical effect if not formal acknowledgement.' (6) Nevertheless, the council remained unhappy about the possibility of unlimited community representation but after a further eight months of negotiations they were unable to get the Home Office to agree to any restrictions.

The issue of community representation does, however, clearly beg the question 'Who decides who the community is?'. The final guidelines attempt to side-step the issue by allowing 'all bona fide, formally constituted bodies which represent a significant number of local people' to be members, (7) but in practice it has been the borough councils which have played a leading role in deciding who participates. In Islington, the council argued for limiting community representatives to twelve and having a balance in favour of ethnic groups and tenants' associations. The police wanted an 'open-ended' arrangement but it was agreed that once nominations had been received a finite limit could be set. A public meeting held to inform the community of the group and receive nominations, was attended by just sixteen people, none of whom represented ethnic groups, most being from tenants' associations. Even after sending nomination forms to all ethnic groups in the borough, the council only received three replies. Informal approaches were then tried by both council and police. The result of these efforts was the attendance at the first meeting of four ethnic groups, five tenants' associations, and representatives of the church, the victim support scheme and the chamber of commerce. There is not the space here to address issues such as the changing representation on the group or whether it is representative of Islington. This sketch is simply to indicate the council's leading role in selecting 'the community' and its attempt to forge alliances with certain sections of the community, the ethnic groups and the council tenants both of which are important in the local political arena.

A similar strategy is apparent in the contrasting political and social environment of Sutton, a relatively prosperous and predominantly white middle-class Conservative-controlled suburb (until May 1986 when the SDP/Liberal Alliance took control) where unemployment is 4 per cent and almost 70 per cent of people are owner-occupiers. (8) Here the council decided unilaterally on the community representatives for the police liaison committee, a third of the places going to chambers of commerce and trade, indicative perhaps of their power and influence in Sutton.

Establishing the police liaison committee in Sutton was

a relatively smooth process, the Council accepting five places on the committee, although insisting that there be sufficient flexibility in the guidelines such that they didn't have to have a Lambeth-style arrangement, a matter of symbolic local importance. Sutton Council wanted the committee to meet in private - 'such a proposal would be quite acceptable here, although it may well not be in Lambeth' - to which the Home Office agreed. At the first meeting, however, the whole committee overturned this decision, deciding to meet in public but reserving part of the agenda for confidental items. This decision was to have important implications, for at a meeting the following year Tory councillors in the public gallery were asked to leave for a confidential item; they asked to stay but the committee refused. This triggered a walk-out by all Tory councillors on the committee, who didn't then return for over a year. In many ways this action represents a similar sense of elitism to that displayed by Islington councillors. At issue was the power and influence of the elected councillors over the unelected police liaison committee and over the police, for they viewed the group as a 'talking shop': 'If criticism was made, a positive response would be expected and although the committee might not be able to give instructions, it was reasonable to expect that it should be able to influence police thinking and attitudes.' (9)

In part this shows a political difference within the Conservative Party between local and central perspectives on consultation: central government, as indicated earlier, wanted to minimize the influence of local government; but in the local political arena councillors are concerned to maintain their autonomy and influence, which were being challenged by the consultative group. The Tory councillors submitted proposals to change the committee's constitution in a way which would make it acceptable to the council but these were rejected by the committee. A Tory councillor commenting on these events said:

> We took the view that we were the only group there that has a mandate from anybody. The others were self-elected from various pressure groups and we did not think that this was a good way to get a representative committee ... I find it sad that we set the darn thing up, put all these people on and they more or less say to us we think we are more important representatives than you.

Nor is this an isolated case of Tory hostility towards consultation. In Hillingdon the council responded to the 1982 guidelines by arguing, as in Islington, that 'consultative arrangements on behalf of the community should be via members of the council'. A Police Liaison Committee was established consisting of five councillors, three council officers, and four police officers; it was closed to the public and the proceedings confidential. Following receipt of the 1984 draft Guidance, the council wrote to the Home Office complaining that 'The latest proposals do not appear to have had regard to experiences in places such as Hillingdon.' (10) At issue was community representation; the council's decision not to have community groups on the committee was because it:

> was convinced that elected councillors would provide a more adequate reflection of local opinion ... (we) can see no justification for elevating community groups to such a position of prominence when it comes to consultation with the police ... There is also a great danger that the proposed arrangement will give a platform to groups which are hostile to the police and result in destructive political conflict rather than constructive debate on the development of policing policies. (10)

This statement echoes the central themes of this section of the chapter: the sense of elitism held by local councillors and the conceptions of 'community' and 'representation' with which they operate. It is these issues which will be of key importance in understanding the local form and process of community-police consultation.

Conclusion

All police forces across England and Wales have been required to make 'arrangements' for obtaining the views of the community about policing since January 1985. A survey of these arrangements has revealed that of the forty-one provincial forces all but three are developing some form of Police Liaison Committee structure (Morgan and Maggs 1985, 25). In London by December 1988 there were only six boroughs where consultative groups still awaited formal approval by the Home Secretary. Preliminary assessments of these groups suggest they provide a forum for the ventilation of local problems and a mechanism for developing practical solutions to some of these problems (Morgan 1987). In Islington, for example, joint police/local authority crime-prevention projects have begun on council estates, and in Sutton the group has produced a set of recommendations to tackle the problem of under-age drinking. More generally, however, the hope expressed in the Home Office guidance that local policing policy could be made more responsive to local needs has yet to be realized. The main determinants of local policing - decisions about priorities, tactics, manpower allocation - are usually fixed centrally at police headquarters and the Home Office. In London, for example, despite concern by consultative groups across the capital about the adverse effects on local policing caused by sending officers to police the News International dispute at Wapping, an average of three hundred officers a day from across the MPD were regularly sent to the picket lines.

This tension between the central and local determinants of policing has emerged as the distinctive feature of the political geography of policing in contemporary Britain (Fyfe forthcoming). Since 1981 there has been a re-structuring of central-local relations which is now gathering momentum. Central government's increasingly interventionist stance is evident both in its general commitment to getting 'value for money' from the police service by issuing guidance on enhancing efficiency, and in its specific involvement in, for example, funding the policing of the miners' dispute which helped to sustain a nationwide police operation. These developments have led to the virtual 'eclipse' of the police authority, the main form of local democratic influence on policing (Spencer 1985). In London much of the momentum of the campaign for an elected police authority has been sapped by the re-election of a Conservative government committed to retaining the current constitutional arrangements. For many Labour borough councils police accountability has become a background issue while crime prevention and community safety are now the policy areas in which these local authorities are trying to expand their role. It is these new developments which provide the substantive basis for developing a political geography of policing and to which a politically informed geography of crime has much to contribute.

Notes

1. Section 106 of the Police and Criminal Evidence Act 1984 requires that 'arrangements shall be made in each police area for obtaining the views of people in that area about matters concerning the policing of the area and for obtaining their co-operation with the police in preventing crime in the area'.

2. For a detailed case study of the Lambeth group see D. Cansdale, 'The development of the Community/Police Consultative Group for Lambeth', unpublished MSc thesis, Cranfield Institute of Technology, 1983.

3. Figures from the 1981 census.

4 Islington Gazette, 11/9/1981.

5. Letter from Chief Executive, Islington Council, to Police Commander.

6. Report of Police Committee Support Unit, September 1983.

7. Home Office, 'Guidance on local consultation', 1985, para. 8.

8. Figures from the 1981 census.

9. Sutton Borough Council, Minutes of Management Committee, October 1983.

10. Letter from Chief Executive, Hillingdon Council, to Home Office, June 1984.

References

Bundred, S. (1982) 'Accountability and the Metropolitan Police' in D. Cowell et al. (eds), Policing the Riots, London: Junction Books, 55-81

Cansdale, D. (1983) 'The development of the Community /Police Consultative Group for Lambeth', unpublished MSc thesis, Cranfield Institute of Technology

Fyfe, N.R. (forthcoming) 'Policing the recession' in J. Mohan (ed.) A Political Geography of Contemporary Britain, London: Macmillan

Greater London Council (1982) The policing aspects of Lord Scarman's Report on the Brixton Disorders, London: Greater London Council

Greater London Council (1983) Policing London, no. 7, April/May, p. 14

Home Office (1985) 'Guidance on arrangements for local consultation between the community and the police in

the Metropolitan Police District, London: HMSO
Lowman, J. (1982) 'Crime, criminal justice policy, and the urban environment' in D.T. Herbert and R.J. Johnston (eds), Geography and the Urban Environment, vol. 5, Chichester: Wiley, 307-41
Maclean, B., Jones, T., and Young, J. (1986) Preliminary Report of the Islington Crime Survey, Centre for Criminology and Police Studies, Middlesex Polytechnic
Mawby, R.I. (1979) Policing the City, Farnborough: Saxon House
Morgan, R. (1985) 'Police accountability: current developments and future prospects', paper presented to the Police Foundation Conference, Harrogate
Morgan, R. (1987) 'The local determinants of policing policy' in P. Willmott (ed.) Policing and the Community, London: Policy Studies Institute, London, 29-44
Morgan, R. and Maggs, C. (1985) Setting the P.A.C.E.: Police community consultation arrangements in England and Wales, Bath: University of Bath
Scarman, Lord (1981) The Brixton Disorders 10-12 April 1981, London: HMSO
Smith, S.J. (1986a) 'Police accountability and local democracy', Area, 18, 2, 99-107
Smith, S.J. (1986b) Crime, Space and Society, Cambridge: Cambridge University Press
Spencer, S. (1985) 'The eclipse of the police authority' in B. Fine and R. Millar (eds) Policing the Miners' Strike, London: Lawrence and Wishart, 34-53
Weatheritt, M. (1986) Innovations in Policing, London: Croom Helm in association with the Police Foundation

Chapter Fourteen

VARIATIONS IN PUNISHMENT IN ENGLAND AND WALES

Linda Harvey and Ken Pease

There is a distinguished psychologist in our university who has spent a large proportion of his professional life doing research into electroencephalography. At times of depression, he is given to complain that trying to understand brain malfunction from electrodes affixed to the skull is a little like trying to discover what is wrong with a car engine from electrodes fixed to the bonnet. Had he chosen criminology as a career, his depression would be much deeper. Attempts to make inferences about the effects of geographic or social variables on punishment patterns is closer to necromancy than electroencephalography. In this brief contribution, we will try simply to describe some of the difficulties, and reach some provisional conclusions about the usefulness or otherwise of the enterprise.

From crime to punishment or from punishment to crime?

It is obvious that people are typically punished because they have committed crimes. Punishments are responses to crimes. Society takes what actions it chooses in defence of itself against criminal depredations. If one conceives of the matter thus, areal and other differences in levels of punishment are appropriately assessed against levels of expectation derived from measures of crime. An area has a high or low level of punishment severity in relation to its level of convictions for non-trivial crime. The use of crime levels as denominators in this way at the least assumes relationships between crime and punishment which are highly problematic. At worst, they misconceive the relationship entirely. First, let us consider the problems with taking a simple-minded view of the 'self-evident' causal order of crime and punishment.

Crimes are not offenders available for punishment

On the basis of simple counting, not all crimes yield offenders to be punished. Most crime is unreported, and this yields no offenders. Most reported crime is uncleared, and this yields no offenders. This is of itself a banal point, but points to the huge untapped reservoir of crime which could be brought into the open in response to the priorities of criminal justice. It allows the possibility that if citizen report and/or detection effort leads to punishment - for instance the courts taking drug possession more seriously - this will be reflected in the level of crime report, the direction depending on whether the actions of the criminal justice process chime with public sentiment. To make the picture still more complex, Chatterton (1983) describes with telling examples how police organizational imperatives work to ensure that many crimes with a low potential for yielding a prisoner remain unrecorded.

A crime recorded brings to light nine

For many crimes the process whereby the offence becomes known to the police is at the same time its clearance. All sexual offences that we can think of generate an offender at the point of the act becoming known. This is likewise true of most drug offences and virtually all fraud. Further, much crime is cleared by means other than detection, notably by offences taken into consideration or cleared by police officers visiting convicted prisoners. The importance of this varies as between police force areas (Burrows and Tarling 1982; Farrington and Dowds 1985). The influence on the decision to record a crime exerted by its potential for generating t.i.cs. (offences taken into consideration) is also commented on by Chatterton (1983).

Public reporting practices change over quite short periods

This is easily demonstrated by reference to changes in the rate of recorded burglary during the 1970s. The absolute increase in the number of recorded burglaries over the decade was of the order of 50 per cent. However, victim surveys incorporated in the General Household Survey in 1972, 1973, 1979, and 1980, and the British Crime Survey whose fieldwork was carried out in 1982, showed that the rate of burglaries sustained over the decade increased only trivially. What appeared to be a burglary epidemic turns out to have been a burglary reporting/recording epidemic, which can plausibly be linked to insurance changes during the decade (Litton and Pease 1984).

In short, the number and type of offenders available for punishment, and serving as the denominator for data on punishment, both depend upon factors in and circumstances surrounding an event which enable classification as a crime. The gearing of the relationships depends upon individual force practices, as was dramatically shown by Farrington and Dowds (1985) in their analysis of differences between practices in neighbouring police force areas in the English Midlands. In the light of this, it would be overly sanguine to expect a coherent geography of punishment to be easily attained.

The pattern of punishments

The raw material for punishment are those people identified as responsible for cleared offences. The relationship between the rate of cleared crime and the number of people processed is crucial. The more modest this relationship, the more criminal justice system factors may be thought to interpose themselves. The simple map of England and Wales depicting the number of people per 100,000 population officially processed for indictable offences (Figure 14.1) cleaves to no obvious demographic or other divisions. It looks crazy, with the Metropolitan Police force area showing lower rates of people processed than Greater Manchester, Nottinghamshire, and Cleveland. The picture circumstantially implicates criminal justice system factors as pre-eminent. This is wholly consistent with detailed analyses in an econometric tradition (Macdonald 1976; Carr-Hill and Stern 1979). Another way of presenting the data is as a scattergram of the relationship between notifiable offences recorded per 100,000 population and the number of people found guilty of or cautioned for indictable offences per 100,000 population. The use of notifiable offences on one axis and indictable offences on the other should make only a slight difference to the relationship. The unit of

Figure 14.1 Persons found guilty of, or cautioned for, indictable offences, per 100,000 population by police force areas in England and Wales, 1986

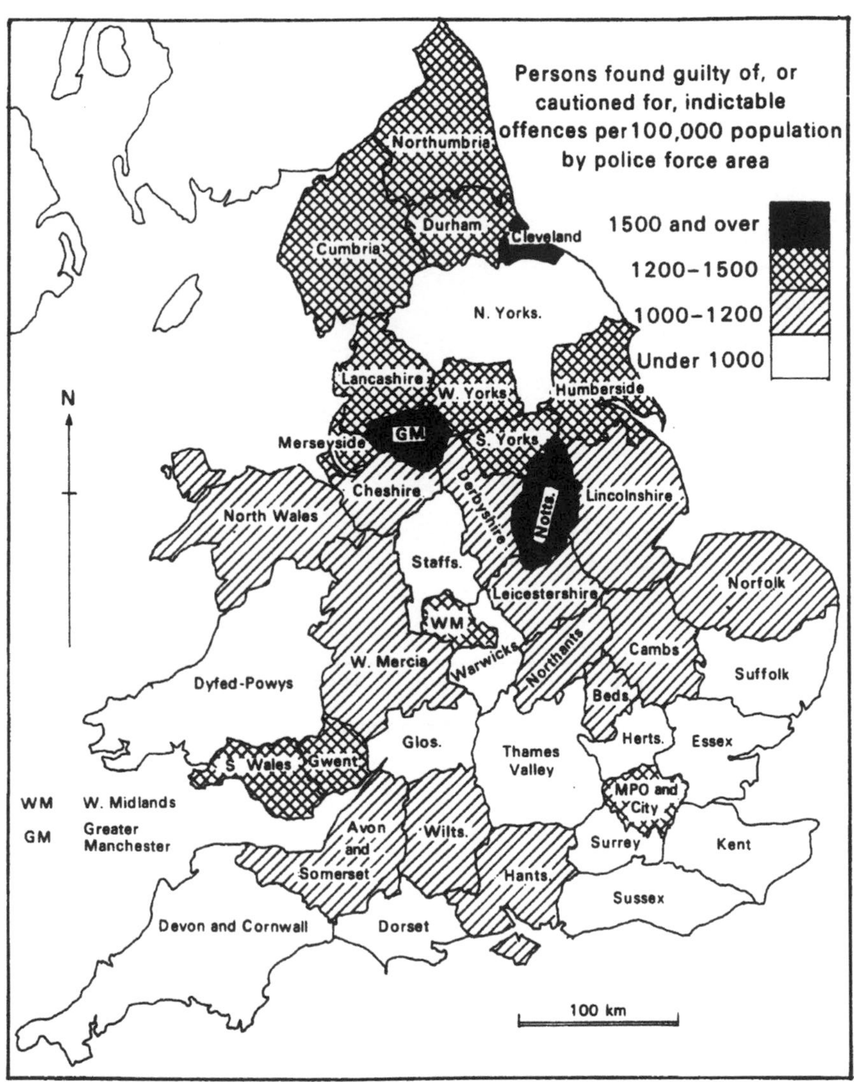

Figure 14.2 Offences recorded by the number of people found guilty of or cautioned for indictable offences per 100,000 population

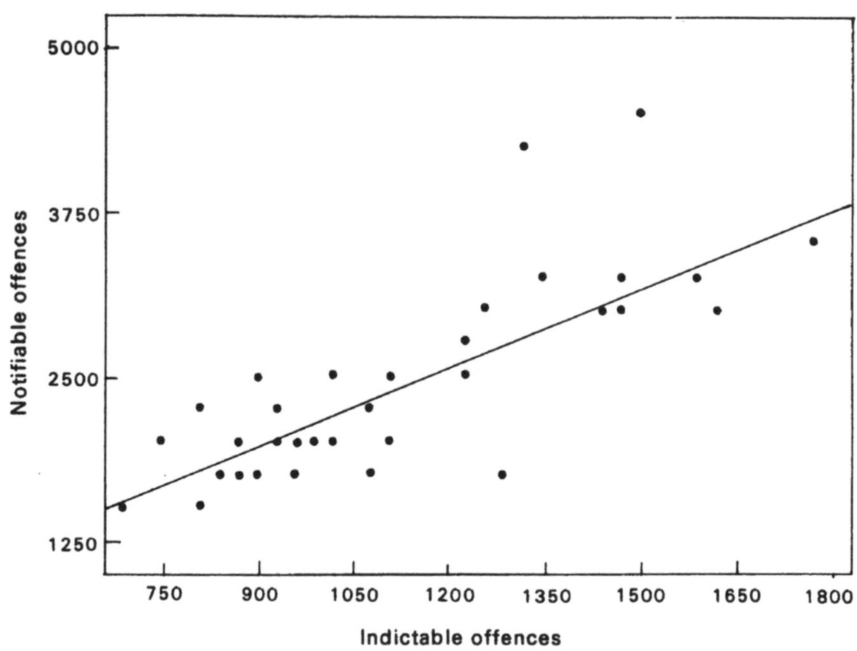

Note: 42 cases plotted. Regression statistics of NOTOFF on INDOFF

Correlation	0.78013	
R Squared	0.60860	
S.E. of Est	453.49852	
Sig	0.00000	
Intercept (S.E.)	-6.73508	(314.74335)
Slope (S.E.)	2.13769	(0.27105)

analysis is the police force area.

It will be noted in Figure 14.2 that the relationship is a modest one (R squared = 0.61). Among the more dramatic outliers are Merseyside and Northumbria. It was stressed earlier that the availability of people for punishment can itself bring to light offences for record. Thus, the relationship between crimes and offenders should not be taken to be causal in the obvious direction. A rough analogy could be drawn with visits to a general practitioner and hospital admissions. The former may be a consequence of the latter as readily as the latter may follow from the former. The important conclusion to which we are drawn is that crimes known to or cleared by the police do not provide a convenient baseline against which to assess differences in punishments, even though, crudely, the first provides the raw material for the latter.

Cautioning rates

The police caution is a pre-court official sanction. It is a formal, recorded warning to an offender by a police officer usually of the rank of inspector. Ever since cautioning ceased to be numerically trivial, there have been substantial area differences in rates of use of the formal caution (see McClintock and Avison 1968 for early data). These have persisted, albeit that they have declined somewhat (Laycock and Tarling 1985). Certain forces consistently appear as high or low cautioners: Bedfordshire, Nottinghamshire, Staffordshire, and Northamptonshire have generally cautioned a high proportion of those eligible, whereas the Metropolitan force, Cumbria, Cleveland, Leicestershire, and South Wales appear consistently towards the bottom of the cautioning league. There is a tendency for rural forces to be characterized by higher rates of cautioning than their urban counterparts, but this is by no means universal. The numbers of cleared crimes per 100,000 population does not co-vary at all with the numbers of offenders cautioned. This argues against seeing cautioning by the police as throwing out junk by a pre-court device.

By what factors then are rates of caution driven? Clearly age and gender are central to the decision. Nationally in 1986, 60 per cent of caution-eligible males between 10 and 16 were cautioned, as compared with 10 per cent of older males. For females the rates are 85 per cent and 24 per cent respectively. Even within age-gender groups, there is wide variation of non-obvious origin between police force areas. For instance Dorset cautions 17 per cent of females aged 17 or over, while Wiltshire cautions 49 per cent. Leicestershire cautions 3 per cent of eligible males aged 17 or over, whereas Nottinghamshire cautions 19 per cent. The general cautioning apparatus, in terms of liaison arrangements and necessary checks and consultations before a caution is issued, appears not to account for differences in rates of cautioning, as a careful reading of Tutt and Giller (1983) makes plain. Nor is there a close relationship between rates of cautioning and the extent to which different police forces proceed against suspected offenders by way of a written summons rather than by arrest and charge (Gemmill and Morgan-Giles 1980).

Another possibility - that the cautioning process anticipates court practice - should be considered, although it is made less likely by the earlier demonstration that forces with high numbers of cleared crimes do not caution more than others. According to this scenario, the police choose not to waste the court's time with the production in court of those offenders whom the court would discharge. This was arguable from the statistics of a decade ago, (Ditchfield 1976), but is so no longer (Bottomley and Pease 1985).

A study by Laycock and Tarling (1985) confirmed the importance of an offender's criminal history for the decision as to caution. Information from nine police forces revealed a fairly close relationship between the proportion of all offenders who were first offenders and the force's cautioning rate. Police forces with higher cautioning rates were also those with the higher proportions of first offenders, and there was much less between-force variation in the rates of cautioning of first offenders. For recidivists, force policy was roughly discernible in force practice. Cautioning was a good place to start. It makes clear that the obvious crime variables do not underpin sanctioning practice in any simple way. The more promising explanatory variables are probably to be found in police force policy differences in cautioning recidivists, and in age and gender differences in cautioning practice and policy. The notion of cautioning practices as police policy driven seems to be the Home Office preference. In its circular 14/1985 the Home Office sets out guidelines on cautioning practice. However, these guidelines are flaccid enough to inspire confidence that reduced dispersion of rates between areas will not eventuate. Indeed the standard deviation of overall force-cautioning rates in 1986 is almost precisely what it had been in 1984, before the guidelines were promulgated (5.526 in 1984, 5.931 in 1986). When broken down by age and gender, however, there are slight differences with higher rates of variation in 1986 for older males and all females:

	1984	1986
Males aged 10-16	7.717	7.522
Males aged 17+	3.585	4.460
Females aged 10-16	5.335	7.333
Females aged 17+	8.661	10.439

Sentencing in magistrates' courts

The vast majority of sentences imposed by courts in England and Wales are imposed by magistrates' courts. Crown Courts get their business through the magistrates' courts. In looking at court differences, we should therefore focus on magistrates' courts, although there is plenty of reason to suppose that there are substantial variations in practice between Crown Court judges (see Wasik and Pease 1986). Many criminologists have looked at variations in sentencing practice. Grunhut (1956) and Hood (1962) are two of the earlier notable British works in the tradition. None of this work has succeeded in accounting for identified variation in terms of characteristics of offenders appearing before the courts. Although its data are now a dozen years old, Tarling's (1979) analysis of sentencing practice in magistrates' courts remains the touchstone for later work. He found that, after controlling for court intake, considerable differences in sentencing practice remained. Notably, differences in probation staffing (which has a regional component) underpinned differences in the use of probation orders. Since probation officers also influence sentencing practice more generally through their recommendations in social enquiry reports, the effect of officer-induced differences will extend beyond probation orders, perhaps particularly to community service orders, where the provision of a court report is a statutory requirement, as is not the case in probation. Tarling (1979) also showed that amounts of fine were linked to social and economic indices (see Table 14.1).

Table 14.1 Correlation coefficients between average fine imposed and social and economic indices

	Average fine	Proportion of males of social class IV and V	Unemploy-ment rate	Average weekly income
Average fine	1			
Proportion of males of social class IV and V	0.41*	1		
Unemployment rate	0.44*	0.42*	1	
Average weekly income	0.30	0.18	0.39*	1

*Significant at 5 per cent level

Despite the very limited predictability of sentencing patterns, the central finding of the Tarling study - let us not forget an officially sponsored study conducted by the now Deputy Head of the Home Office Research and Planning Unit - was that courts even within a police force area were not consistent in their imposition of sentence and were in fact not unduly concerned to achieve such consistency. For instance, for motoring offences the Magistrates' Association issues a list of recommended penalties. While the proportion of courts asserting fealty to such a list has grown (compare Hood 1972 with Tarling 1979), Tarling found adherence to it to be limited. Table 14.2 shows the degree of adherence.

Even locally, consultation seemed to be minimal. As Tarling commented:

> While nearly all courts claimed to have worked from the Magistrates' Association list in setting their own basic norms, only about half of them consulted their neighbours in doing so. Even then consultation rarely accounted for more than an exchange of penalty lists and it was unusual for all courts within an area (e.g. all those within an area served by one magistrates' courts committee) to meet together to agree on the penalties to be imposed in that area. Only four courts had done this (p. 32).

Table 14.2 Variations in recommended penalties for various types of motoring offences between twenty-one magistrates' courts

	Lowest recom-mended fine	Highest recom-mended fine	Fine recom-mended by Magistrates' Association	Number of courts using Magistrates' Association fine +/- 10
Drunken driving	50	100	80	12
Driving while disqualified	50	100	100	12
Dangerous driving*	50	100	100	8
Careless driving	25	50	50	9
No insurance	30	100	50	18
Taking and driving away a motor vehicle*	40	100	60	12

* One court omitted dangerous driving and four courts taking and driving away a motor vehicle from their lists.

The more general conclusion about the lack of concern with cross-court consistency merits quotation at length.

> (Courts) were really only interested in maintaining consistency within their own practice. With the exception of trying to even out the levels of fines for motoring offences within a locality, no court spokesman claimed that any attempt was made to achieve consistency with his neighbour or with courts over a wider area or with what was thought to be national practice. None of the courts knew much about the sentencing practice of any others, even about those next to them, and they regarded such knowledge as irrelevant to their own decisions or problems. What comparisons were made were usually little more than anecdotal. Occasionally a clerk or magistrate spoke of his own or a neighbouring court's reputation for severity or leniency, but in general it was thought that differences in the sorts of crimes dealt with by different courts precluded any useful comparisons

> between their sentencing policies. Rightly or wrongly, local conditions and patterns of crime were seen as peculiar to that locality, and sentencing policy as thus to be uniquely adapted accordingly. This conviction on the part of influential figures in the courts, together with a strong and general tendency towards the preservation of autonomy, seems to ensure that very little emphasis is placed on action to achieve anything like a more uniform approach to sentencing even within a fairly small geographical area.
>
> (Tarling 1979, p. 27)

Magistrates' sentencing and employment variation

There is a relationship between being unemployed and committing crime (Farrington et al. 1986). Over and above that, there is a putative relationship between being unemployed and being sentenced severely. Farrington and Morris (1983) in a study of sentencing in a single magistrates' court found that being unemployed was associated with receipt of a more severe sentence. In a NACRO review (1983), it was contended that the link between a country's unemployment rate and its prison population could not be accounted for by variations in crime rate.

In the most sophisticated study of the relationship carried out to date, Crow and Simon (1987) tried to separate out the variables of traditional attitudes to work and extant sentencing practice. Six courts were selected. Two were in areas with a continuing high rate of employment. Two more were in areas where the opposite was true. The final two were in areas where unemployment had become a major problem only in recent years. Within each pair of courts, one had a high rate of custodial sentencing, the other a low rate. It was thus possible to look at differences in sentencing in magistrates' courts according to experience in relation to the labour market and according to the court sentencing tradition. It becomes clear in Crow and Simon's data that their design was well-founded, since both factors are of relevance in accounting for the observed pattern of relationships after offence and criminal history are controlled for. Specifically, for men aged 21 or over, there was a significantly higher use of immediate imprisonment in those courts with a traditionally high use of custody when levels of unemployment were low or recently increased, that is, when they could be interpreted as characteristics of the offender rather than circumstances.

These differences were neither so consistent nor so marked for offenders under twenty-one. The employed were more likely to be fined in both age groups and for almost all courts. Crow and Simon comment 'Offending score being allowed for, the employed were more likely to be fined, whereas for the unemployed there was a tendency to displacement, some moving down to conditional discharge ... and others moving up towards higher tariff disposals' (p. 23).

The particular contribution of the NACRO study is that it highlights the observation that area differences should properly be thought of as complex and partial determinants of sentencing patterns. There are no simple crime-based determinants of them.

Court differences within a police force area

It has been suggested earlier that rates of cleared crime do not account for variation in sentencing. It also seems clear that rates of cautioning stem more from police force policy and offender age and gender than more crime-related variables. Attempts to control area autonomy in cautioning practice seem feeble.

The rhetoric of sentencing is a rhetoric of autonomy. Area differences here are at least as likely to be the product of caprice as is the case for cautioning. In the discussion of what topic of social concern other than sentencing we would take seriously, it could be argued that area differences are the result of undemonstrated differences between towns; there may be some measure of uncontrolled discretion in decision-making. Using research on the relationship between employment status and offending as an example, one can discern that court tradition and the 'meaning' of unemployment for the sentencer in her presumption of the character of the offender may well underlie sentencing differences.

The last piece of evidence which will be brought into play concerns differences in sentencing practice within police force areas. If different court areas lie within the same police force area, police force policy cannot be the reason for different sentencing practices. The commonsense prediction would be that court propinquity and sentencing similarity should go together. If they did, the reason would

Table 14.3 Percentage of all magistrates' court disposals which are custodial, for males aged 17-21, in Greater Manchester, 1986

Division	Percentage
Manchester	12
Oldham	13
Middleton	18
Rochdale	12
Eccles	17
Salford	20
Stockport	16
Ashton	23
Sth Tameside	24
Trafford	18
Leigh	11
Makerfield	17
Wigan	12

Source
Figures obtained from: Table S5.9, Criminal Statistics England and Wales, Supplementary Tables, 1986, vol 5 (HMSO).

be obscure. It could result from greater similarity of closer courts along social variables, or greater influence of close courts, one upon the other. Further analysis would be necessary to untangle the alternatives. If, on the other hand, there is no greater similarity between adjacent than between non-adjacent courts, sentencing courts can be regarded as islands, entire unto themselves.

The datum discussed here and illustrated in Table 14.3 is the proportion of all magistrates' court disposals which are custodial (this includes partly suspended sentences and committals to the Crown Court, but not wholly suspended sentences). As Table 14.3 shows, areas of Greater Manchester with very similar proportions of custodial disposals are not necessarily close in geographical terms. Indeed, Manchester and Salford, which are adjacent geographically, have custodial disposals for young males which are 12 per cent and 20 per cent of total disposals respectively. On the other hand, Stockport and Eccles, which are on opposite sides of the conurbation, have similar levels of custodial disposal (16 per cent and 17 per cent respectively).

It seems that adjacent courts' sentencing practice, at least in respect of custody, is not more similar than the practice of courts which are further apart. In this, factors other than local similarity and influence appear to be central. Fiercely protected local autonomy rules, together, by common consent, with the rule of that -minence grise of the criminal justice system, the Court Clerk.

Criminal justice differences

None of the approaches outlined above unequivocally demonstrates the insulation of court sentencing practice from crime variables. However, taken together they certainly suggest that the source of criminal justice differences are primarily differences in the perceptions and practices of agents of criminal justice. So what?

Modelling of the determinants of crime variation is defensible. It can aid the allocation of criminal justice resources and the planning of places and social arrangements in ways which are less criminogenic. (Although one would be able to count the number of cases where this has happened on the fingers of one badly-mutilated hand). The same argument of applicability does not so easily apply to the modelling of punishment processes. Wherever we look, it seems that punishment decisions are already more a function of policy than the fund of punishable persons. Punishment policy choices have consequences for rates of crime, and the analysis of individual careers is already providing data which can inform punishment choices (Farrington 1987). Punishment patterns are also defensible or otherwise in terms of morality. This has spawned a literature too extensive to need referencing. In short, punishment patterns are already a product of policy, typically local policy. They always must be so. No conceivable demonstration of how one or other variable impacts on punishment would render that untrue. Punishment is almost always best conceived as an independent variable. Analysis of punishment variation may depict a situation of which to disapprove, but is otherwise useless. The policy choices remain the same whatever the outcome.

A future

If current area differences in punishment are regretted, there is a possible touchstone available in local public opinion. Surveys of victimization typically elicit views of punishment, and when samples become usable by aggregation over time, as is the case for example in the successive runs of the British Crime Survey, enough data will be generated to provide some insight into local punishment preferences. Early results along these lines already evoke surprise (Hough and Moxon 1985). Questions must be refined to reflect financial factors. The opportunity costs of expensive punishment should be made clear. However, the basic approach has the merit of cutting the Gordian knot, of feeding public punishment preferences directly into the consciousness of sentencers, rather than by the processes of the press and preconceptions about what people think and want.

Acknowledgements

ESRC help to the first author during the preparation of this paper is gratefully acknowledged.

References

Bottomley, A.K. and Pease, K. (1985) Crime and Punishment: Interpreting the Data, Milton Keynes: Open University Press

Burrows, J. and Tarling, R. (1982) Clearing Up Crime, Home Office Research Study No. 73, London: HMSO

Carr-Hill, R.A. and Stern, N.H. (1979) Crime, the Police, and Criminal Statistics, London: Academic Press

Chatterton, M.R. (1983) 'Police in social control', in J.F.S. King (ed.) Control Without Custody? Cambridge: Institute of Criminology, University of Cambridge

Criminal Statistics England and Wales 1986, (1987) Cmnd 233

Criminal Statistics England and Wales, Supplementary Tables 1986, vol. 5, (1987) London: Home Office

Crow, I. and Simon, F. (1987) Unemployment and Magistrates' Courts, London: NACRO

Ditchfield, J.A. (1976) Police Cautioning in England and

Wales, Home Office Research Study No. 37, London: HMSO

Farrington, D.F. (1987) 'Predicting individual crime rates', in D.M. Gottfredson and M. Tonry (eds) *Crime and Justice*, 9, 53-102

Farrington, D.P. and Dowds E.A. (1985) 'Disentangling criminal behaviour and police reaction', in D.P. Farrington and J. Gunn (eds) *Reactions to Crime: the Police, Courts and Prisons*, Chichester: Wiley

Farrington, D.P. and Morris, A.M. (1983) 'Sex, sentencing, and reconviction', *British Journal of Criminology*, 23, 229-48

Farrington, D.P. *et al.* (1986) 'Unemployment, school leaving, and crime', *British Journal of Criminology*, 24, 335-56

Gemmill. R. and Morgan-Giles, R.F. (1980) *Arrest, Charge and Summons: Current Practice and Resource Implications*, Royal Commission on Criminal Procedure, Research Study No. 9, London: HMSO

Grunhut, M. (1956) *Juvenile Offenders before the Courts*, Oxford: Clarendon Press

Hood, R. (1962) *Sentencing in Magistrates' Courts*, London: Stevens

Hood, R. (1972) *Sentencing the Motoring Offender*, London: Heinemann

Hough, J.M. and Moxon, D. (1985) 'Public attitudes to sentencing offenders', *Howard Journal of Criminal Justice*, 24, 93-112

Laycock, G.K. and Tarling, R. (1985) 'Police force cautioning: policy and practice', *Howard Journal of Criminal Justice*, 24, 81-92

Litton, R.A. and Pease, K. (1984) 'Crimes and claims: the case of burglary insurance', in R.V.G. Clarke and T.J. Hope (eds) *Coping with Burglary*, Lancaster: Kluwer-Nijhoff

McClintock, F.H. and Avison, N.H. (1968) *Crime in England and Wales*, London: Heinemann

Macdonald, L. (1976) *The Sociology of Law and Order*, London: Faber and Faber

NACRO (1983) *Unemployment and Imprisonment*, briefing paper, available from NACRO, 169 Clapham Rd, London SW9 0PU

Tarling, R. (1979) *Sentencing Practice in Magistrates' Courts*, Home Office Research Study No. 56, London: HMSO

Tutt, N. and Giller, H. (1983) 'Police cautioning of juveniles: the practice of diversity', Criminal Law Review, 587-94
Wasik, M. and Pease, K. (1986) Sentencing Reform, Manchester: Manchester University Press

Chapter Fifteen

CRIME PREVENTION: THE BRITISH EXPERIENCE

Gloria Laycock and Kevin Heal

The prevention of crime is currently a subject of considerable interest and activity within this country and abroad (Heal and Laycock 1986) But the artefacts of pre-history suggest that property marking - an activity now much in favour - was known and used by paleolithic man to protect his few possessions. The historian's commentary on the medieval castle, the practice of 'hue and cry' and the Elizabethan fortified manor house, show beyond doubt that current interests in target hardening, community surveillance and design are not new.

These long-standing preventive traditions were reinforced, as far as the police were concerned, by the Metropolitan Police Act 1829. Nevertheless, by the middle of this century the police service had become largely reactive, devoting its resources primarily to investigation, detection, and prosecution. The 1960s saw an important attempt by central government to change the balance - to move from declaratory support for prevention to tangible support in practice. In 1960 the then Home Secretary set up a committee on the prevention and detection of crime under the chairmanship of W.H. Cornish, a senior Home Office official. The committee examined a number of issues but gave particular attention to crime prevention and the role played by the police. The present Home Office Standing Conference on Crime Prevention (described in more detail later) sprang from amid its 117 recommendations, as did the post of police crime-prevention officer, central crime-prevention training and, in embryonic form, the crime-prevention panel. The committee thus laid the foundations of a structure for crime prevention still visible today.

Unfortunately these attempts to foster prevention were largely overtaken by other contemporary changes within policing which led to a massive expansion in vehicular patrolling, a development which reduced contact between the police and public, and the growth of information technology which, in the first instance, was deployed to reduce police response time - the benchmark of reactive policing.

By the mid 1980s a police service of over 221,000 men fielded no more than 600 crime-prevention officers. Prevention remained the poor relation within the organization. This view can of course be challenged by the argument that the police officer patrolling the streets also prevents crime, but there are severe limitations on the extent to whch this is so (Clarke and Hough 1984).

It is against a background of growing individual, and to some extent community interest in the prevention of crime, but little by way of organizational support, that this chapter looks at some of the developments in crime prevention in England and Wales over the last ten years. It is, therefore, an essay in contemporary history but one which, while reviewing the past, also looks to the future and the difficulties to be overcome if prevention is to secure a more permanent place in society's response to crime.

The renaissance of crime prevention: the catalyst for change

Few social policies have a single root and the change now bearing on crime prevention is no exception. By the end of the 1970s several influences were coming together which were to result in prevention being given greater weight; the strain on the formal criminal justice system (courts, police, probation, and prisons) was starting to become apparent. Despite the very considerable efforts of those involved, and the increase in resources deployed, recorded crime continued to grow at somewhere between 5 per cent and 7 per cent per annum, a trend reflected in the figures of the past thirty years. The cost of maintaining the criminal justice system accelerated, while the prison population reached what many regarded as an unacceptable level, and detection rates fell. Under these circumstances a 'more of the same' solution was clearly no longer tenable, and the findings of the Cornish Committee had to be rediscovered.

Research reinforced this gloomy picture by pointing to

the limited effectiveness of policing and sentencing as measures to control crime. But alongside this work was a more constructive note. Research activity focusing on crime patterns, which subsequently led to the first British Crime Survey (Hough and Mayhew 1983), started to provide a more detailed picture of crime. It became widely accepted, for example, that recorded crime figures should be seen only as the tip of an iceberg of criminal activity since much crime was unknown to the police. A natural consequence of this was to alter the balance in favour of preventive measures since, by definition, reactive tactics could have no impact on unreported crime - the extent of which was clearly considerable. Research also reinforced and substantiated the anecdotal evidence of the Cornish report that much crime was casual and could readily be prevented by blocking opportunity. Starting with burglary and autocrime, work by the Home Office Research and Planning Unit, the Royal Canadian Mounted Police, and the Swedish National Council on Crime Prevention illustrated that many aspects of crime were related to environmental design. This message directed attention to the potential contribution to crime control from planners, architects and local authorities, i.e. professional groups and agencies far removed from the criminal justice system.

Against this flood of information it was no longer acceptable to deal with crime as an occasional ad hoc disaster calling for a specific response. Crime came to be seen as a part of the fabric of society occurring where opportunities were greatest. The criminal justice system, rather than being the cutting edge against crime, became an activity of last resort. The main thrust for crime-control policies was seen to rest with those who shaped communities, for example, the local authority; and with those who comprised communities, the public. It was this view that opened the door to prevention and provided a logical argument for sharing responsibility across a range of agencies and individuals. Public expression of these views came in March 1982. During the course of a debate in the House of Lords the Lord Chief Justice said:

> Neither police, nor courts, nor prisons can solve the problem of the rising crime rate. By the time he (the accused) reaches the court it is too late. The damage has been done. The remedy, if it is to be found, must be sought a good deal earlier.

On the following day the then Home Secretary, Viscount Whitelaw, stated in the House of Commons that a realistic strategy against crime, as opposed to a strategy for the criminal justice system, must recognize that crime is a problem for all of the community. He reported his agreement with ministerial colleagues that, quite separate from direct police action, there was considerable scope for local initiatives to reduce opportunities for crime, and for the encouragement of a sensible climate of opinion and local leadership. This Ministerial statement led to a series of initiatives including: the formation of a group to consider the government's response to crime with membership drawn from several central government departments under the chairmanship of the Permanent Secretary of the Home Office; a major seminar on crime prevention bringing together central and local government officials, the police, and the voluntary sector; and, in January 1984, the preparation of the joint circular (Home Office Circular 8/84) on crime prevention. This activity reinforced the need for a collaborative response, while the notion of opportunity reduction provided a lead for the practitioner wishing to devise effective preventive action. The preliminary outline of a policy, which research had suggested might be successful, was beginning to take shape. Moreover it was a policy which, when set against current conventional wisdom that 'nothing works' carried with it an unusual note of optimism.

Activity at national level

From 1982 onward activity at national level grew steadily. In the broadest terms the objective of this work was to ensure that, where relevant, mainstream social policies, services, and programmes contributed to crime prevention. There was seen to be scope particularly in the areas of housing, education, social security provision, transport, and private sector involvement. In promoting the idea of prevention the Home Office worked in collaboration with other central government departments and voluntary organizations, the police, some local authorities, and the probation service. Some of the work done by these organizations is discussed below.

Recent central government activity

In 1983 the Home Office Standing Conference on Crime Prevention (established following the Cornish Committee) was reconstituted and strengthened by a Home Office Minister taking the chair. Also in 1983 the Home Office Crime Prevention Unit was set up with a specific remit to operate beyond the Office in promoting crime prevention. In 1986 an inter-departmental Ministerial Group on Crime Prevention was inaugurated and two seminars on crime prevention were held at 10 Downing Street, the first under the chairmanship of the Prime Minister, the second led by the Home Secretary.

Other government departments have also become more heavily involved, often through the work of the Ministerial Group. For example, in 1986 £15 million was provided by the Department of Transport to London Underground Limited for work on the reduction of crime and disorder on the underground system, this to involve changes in the layout and design of stations and revision of management procedures.

In looking at the incidence of domestic burglary on local authority housing estates it became clear that a substantial number of offences involved theft from pre-payment coin meters (Hill 1986). Action has been taken to reduce the number of these meters on the vulnerable council property in which they tended to be concentrated. This has required the Department of Energy to work in collaboration with the fuel supply industry, and has led to the speeding up of the programme of replacing existing meters by 'cashless' or token meters.

The Department of Health and Social Security has expressed a growing interest in reducing the crime associated with hospital and other health service premises, while the Department of Education and Science is supporting the development of social responsibility courses in schools which will include schemes aimed at preventing crime.

Other than the Home Office with its contribution through policing, the Department of the Environment remains the government department with the most significant part to play in the prevention of crime - through environmental change. The control exercised over the design, layout, and management of housing estates is potentially of major significance as a means of reducing crime and misbehaviour. Whilst the argument over the precise nature of the part played by design in crime control remains (see for example, Poyner 1983; Coleman 1985; Hope 1986), there seems to be little doubt that housing allocation policies on the broader front are crucial if problem areas for the police, with all that this implies for social breakdown and crime, are to be avoided (Bottoms and Wiles, forthcoming). The Department of the Environment's Priority Estates Projects are also potentially valuable in reducing crime with their emphasis on improved housing management at the local level (Burbidge 1981).

The police service

The emphasis on crime prevention nationally has also recently been reflected in changes within the police service. The staff and resources of the Home Office Crime Prevention Training Centre at Stafford have been increased and the curriculum substantially revised, moving away from the previous 'locks and bars' emphasis towards community involvement, crime pattern analysis, and inter-agency work. The status of the centre has also been enhanced within the police service by raising the rank of its director from Chief Superintendent to Assistant Chief Constable.

A more significant development was the decision to form, within the committee structure of the Association of Chief Officers of Police (ACPO), a sub-committee dealing solely with crime-prevention matters and reporting directly to ACPO's main Crime Committee.

A practical consequence of these developments is the growing recognition that all police officers (uniform and CID) can contribute to crime prevention provided they are given the necessary organizational support and training. There is a steady shift, therefore, away from the view that day-to-day crime-prevention advice should only be offered by staff from a specialized department isolated from the mainstream of policing. The task is being pushed down to the officer on the beat while the crime prevention officer is increasingly becoming a more valued resource within his force for other activities. For example, in some areas the force crime-prevention officer is taking a 'problem-solving approach' to prevention in which the local crime data are analysed on the basis of which a preventive strategy is developed. In other areas the police are becoming a catalyst for change within the community: a person assisting others to pursue prevention perhaps by leading discussions with the local authority, transport undertakings, or local commercial groups. Some of these activities are discussed in more detail in a later section of this chapter.

Private-sector involvement

In some respects the changes which have occurred in central government and the police might have been expected, following the flurry of activity in the early 1980s. What is perhaps more significant are the notable changes within the private sector. Some manufacturers are, for example, now taking account of security at the design stage of production. Some private vehicles are now produced with better door and window locks than hitherto, and a small number of builders are taking account of the need for improved domestic security in building residential property. This activity is being supported by the British Standards Institute and the National House Building Council.

From the outset it has been recognized that simple changes in design will not be sufficient to reduce crime unless accompanied by good management practice in, for example, organizations such as banks, building societies, licensed premises, schools and retail outlets. Recent reports on crime associated with licensed premises, shops, and commercial organizations handling large sums of cash (Home Office 1986a) point to a growing recognition amongst the private sector of the steps that can be taken to reduce the vulnerability of their premises and their staff.

There is also some indication that the private-sector companies are beginning to turn their enormous advertising budget in support of prevention. Some car manufacturers are now encouraging the public to think about the security of the vehicles they are buying as well as their safety or comfort. Other companies, looking to the welfare of their own employees, are advertising the need for domestic security in house journals.

Lastly, some insurance companies have revised the structure of their premiums to ensure that policy holders either benefit financially from taking preventive action (such as the installation of window and door locks or belonging to a neighbourhood watch scheme), or are penalized (through the introduction of financial excess) because they have chosen not to protect their property. Sums of money involved are at present small, but again a principle has been established which five years ago would not have been recognized.

The effect of national activity

The products of this activity are not easy to classify and assess. For example while there has been a steady stream of guidelines, research papers, seminars and conferences, and much inter-departmental activity, it is difficult to be sure whether the general public's awareness of prevention has increased, and if it has, whether this has been translated into sensible practical action. The notion that crime and its prevention is the responsibility of the police and the criminal justice system is entrenched; it is hardly going to be turned round overnight. Nevertheless there are some signs that responsibility is being accepted by organizations beyond the police; some local authorities, for example, are beginning to see that they have a major contribution to the prevention of crime and to the quality of life more widely; some are actively setting about the establishment of a corporate crime-prevention policy.

Activity at local level

It is unrealistic to assume that activity at national level alone will be successful in curbing or reducing the problems associated with crime and misbehaviour in society. While in certain respects central activity is extremely valuable, (for example, in providing national policies and guidelines which, in their formulation, are designed to combat crime), the fact that crime patterns vary from area to area and frequently from street to street, means that national action must be complemented by practical, locally based initiatives. It is in the community that preventive initiatives can be best devised and turned to maximum benefit. It is also in the community that the individual can play a part in reducing crime and so gain the confidence that he is not powerless in its face. As the following paragraphs show, the variation in objectives, organization, and impact of local crime schemes is considerable, and it is probably to the advantage of crime prevention that this is so.

Local police activity

One of the more important changes as far as local police activity is concerned is the provision of crime-prevention advice to the public. This is no longer regarded as the sole responsibility of the force crime-prevention officer. Increasingly foot and vehicle patrol officers, detectives investigating crimes, community constables, and home beat officers are, in many forces, the first point of contact for the member of the public seeking information. As a consequence the interface between police and public on crime-prevention matters is considerably broader than hitherto.

There have been changes in other areas, for example, in the work of crime-prevention panels. These locally-based groups of members of the public have been in existence for over twenty years in some parts of the country. Traditionally they were chaired by a police officer and met perhaps quarterly to discuss local crime problems. With some notable exceptions, this may have led to a leaflet campaign for the elderly or some other publicity-based activity but it has generally been rather unfocused and difficult to evaluate as a consequence. A national crime-prevention panels conference held in 1984 brought new energy to the movement and a paper produced by the Home Office Crime Prevention Unit pointed to the need for changing the role of crime-prevention panels to take account of recent developments (Smith and Laycock 1985). The role and method of operation of panels was more comprehensively reviewed in 1985 by one of the Home Office Standing Conference Working Groups (Home Office 1985) whose report led to the introduction of new guidelines for the operation of panels. One of the more important consequences of this activity (in the interests of drawing members of the community into the planning of crime-prevention work) has been the decision in many forces that the chairperson of local panels should be a civilian rather than a police officer as was the case in the past. While the police continue to attend crime-prevention panel meetings, their role is increasingly becoming supportive rather than directive.

The police have assumed a similar role in respect of neighbourhood watch. Community demand has driven the number of schemes from 1 in 1982 to over 35,000 in 1987, an expansion which illustrates the extent of community resources which can be deployed against crime. The growth of neighbourhood watch has, of course, brought problems for the police, not the least of which is the demand schemes make on police resources. There is a particular complication here since for the most part schemes tend to flourish in the more tranquil parts of the country so, potentially, syphoning police manpower away from hard-pressed areas. Understandably, therefore, there is a growing interest in 'low-energy' schemes now being established in some areas, which minimize police input, and in attempts to locate embryonic schemes in high-crime areas. In practice a second generation of schemes is now being established to cope with the problems identified following the massive expansion of this form of crime-prevention activity.

Local voluntary work

The most active of the voluntary agencies in crime prevention is the National Association for the Care and Resettlement of Offenders (NACRO). In the mid 1970s, with the full support of the local police, NACRO launched a pilot project at Cunningham Road in Halton, Cheshire. This initiative marked, for many, the beginning of the neighbourhood approach to crime prevention, an approach which offered a project development service to local authorities, tenants' groups, and other bodies concerned to reduce crime problems on housing estates. The scale of the work increased steadily to the point that by January 1987 over 80 local projects had been established (Shapland and Osborn 1987).

More recently NACRO have begun to develop the notion of 'community safety', in preference to crime prevention. A key feature of community safety is the development of a unified policy tailored to the needs of the particular area which is more in keeping with what the individual members of the community see as appropriate. This takes on a wider remit for action than tackling the crime level alone and addresses fear of crime, the safety of vulnerable groups such as the elderly, racial minorities, or women, support for victims, and the behaviour and treatment of young people.

Central government at local level

The late 1970s saw a scheme similar to NACRO's Cunningham Road Project established by the Department of the Environment (DOE) through their Urban Housing Renewal Unit, under the title of the Priority Estates Project (PEP). Here the aim was to give local authorities guidance in improving the management of difficult estates. While not specifically aimed at crime prevention many of the techniques recommended to local authorities had preventive implications, particularly PEP's interest in tenant participation and consultation, and the establishment on site of housing management and repair services and, where appropriate, the formation of special project teams. The crime element in PEP's work arose from the tenant consultation process since, not surprisingly, crime problems - particularly vandalism, burglary, fear and anti-social behaviour - were frequently among the issues raised by tenants and housing officers of the estates where the PEP teams were working. As a result many of the improvements introduced through PEP included entryphone systems, door porters, and better lighting, as well as improved management and repairs services.

It is clear, therefore, that by the time the major expansion in crime prevention took place in the early 1980s local projects were well established at least in some parts of the country, and some of the lessons associated with their introduction had been learned. However, the new round of activity brought with it further local action in the form of the MSC Community Crime Prevention Initiative. The Community Programme is a temporary employment scheme, operating throughout Britain and organized by the Department of Employment, which brings together people who need work and have been unemployed for some time with the work which needs to be done - work of practical benefit to the community. A wide variety of organizations such as local authorities, voluntary bodies and private companies run schemes under the Community Programme. Funding for projects is currently limited to wages, based on local authority pay scales, for managers and supervisors (normally on a 1:10 ratio of managers to participants) and participant wages averaging £67 a week.

In 1987, 8,000 job places under the community programme were specifically allocated to crime-prevention work. This involves either working to protect people or property or working to develop social and community activities. A recent publication (Home Office 1986b) provides a number of case studies as examples of successful crime-prevention schemes operating under the community programme. The key features of these schemes are that they:

(i) Should aim to tackle local problems, locally identified
(ii) Should be supported by local people who should, if possible, be a part of them
(iii) Should involve local organizations, such as local authorities and voluntary bodies in addition to the police
(iv) Should be promoted and publicized with sensitivity and care

The case studies include examples of lock-fitting schemes for old and disadvantaged people; improved management of tower block housing; the provision of support for women victims of domestic violence; providing youth activities in disadvantaged areas; and the provision of crime-prevention advice and publicity material.

In addition to its direct impact on crime through improved security and greater awareness of the importance for the individual of taking precautions against crime, all of this work involves young people who may be at risk of committing offences themselves. It thus aims to serve crime prevention in two ways - by making crimes more difficult to commit and by offering alternatives to criminal behaviour either through the work itself or by providing increased community opportunities for leisure pursuits.

A separate initiative, but one drawing in part upon the MSC Crime Prevention Programme followed from a series of requests from local authorities, the police, and voluntary organizations for guidance on establishing crime-prevention schemes. In October 1985 the Home Office announced its decision to set up 'demonstration projects' in five towns. By January 1986 five towns had been selected and activity was under way. The five projects are located in Bolton and North Tyneside in the North of England, Wellingborough in the Midlands, Croydon in the South of England and Swansea in Wales. The aim of the initiative, which was funded for 18 months by central and local government, was to generate public confidence that crime and the fear of crime could be reduced. Each initiative comprised a number of individual schemes requiring the involvement of local agencies and voluntary bodies, and of course, the local residents themselves. The schemes were established on the basis of a detailed area crime profile - drawn from statistics provided by the police and other agencies - which highlighted which particular local crimes were most frequent or caused most concern.

In each area a local steering committee was established, made up of representatives from interested local organizations and this helped with the design of local schemes and the implementation of agreed measures. The projects are now drawing to a close and final reports are expected shortly.

Review of local activity

One of the more obvious features in the current local crime-prevention scene in England and Wales is its fragmented nature. For some, the absence of a systematic structure across the country, springing from central government and based perhaps on a local authority network, is seen as a mark of failure. Those supporting this view look to the continent for examples of good organizational practice (cf. developments in France). Clearly a systematic structured approach has advantages, particularly in giving central departments a greater hand in the development of prevention. To be set against this, the more fragmented approach found in England and Wales permits considerably greater flexibility from one area to the next, so allowing local needs to be met and the unique characteristics of each area to be harnessed to preventive action. Moreover, it is an approach which allows maximum use of existing regionally and locally established networks. Already the CBI, TUC, and ABCC are starting to work to the benefit of crime prevention, and in doing so are starting to take possession of the concept of prevention.

Lastly, it is an approach which readily allows grassroots movements to break through in a way which would undoubtedly be more difficult in a situation where a mono, as opposed to a pluralistic organizational approach to crime prevention has been established. The growth of neighbourhood watch, a development which to date (at the most conservative estimate) has caught the imagination and practical support of around two million people, is the clearest example of this.

Future trends

Looking to the immediate future, where does crime prevention appear to be moving? One might expect to see the contribution from local organizations growing, hopefully with the support of improved information on crime patterns. Techniques are being developed (Houghton and Berry 1987; Ekblom 1987) which will encourage the local area analysis of police crime data and lead to the development of targeted crime-prevention schemes.

Local activity will also be strengthened by the development of networks bringing together people with similar interests and engaged in similar activities. The 1984 National Crime Prevention Panels Conference, and subsequent regional panel activity, provided a framework for the work of these groups. Similar activity is also likely in respect of MSC Community Programme schemes and, of course, neighbourhood watch. The decision by the Home Secretary to explore the possibility of a National Organization for Crime Prevention (NOCP) may well mark a further turning point. If established, the NOCP would not only draw together local schemes, but also provide a much-needed independent voice for prevention.

Developments at local level will almost certainly be matched by activity within central government giving a greater co-ordination of departmental policies. In this, the strategy against crime, as opposed to the strategy for the criminal justice system envisaged by Viscount Whitelaw, will become a reality.

Before this can be achieved, however, it will be necessary to have a far clearer idea as to how mainstream departmental policies in the areas of health, housing, social security, and education, for example, bear on and shape crime patterns at local level. This will call for a major step forward in crime-prevention thinking and a far clearer picture of crime than exists at present. At a time when policy-makers, administrators, and practitioners are moving swiftly to strengthen organizational machinery for prevention at national and local level, to exploit existing networks, and to develop new ones for the benefit of prevention, what is in doubt is the ability of researchers, policy analysts, and criminologists to keep pace and to provide sufficiently precise information about crime and criminality to justify the machinery's existence.

References

Bottoms, A.E. and Wiles, P. (forthcoming), 'Crime and housing policy: a framework for crime prevention analysis' in Hope, T. and Shaw, M. (eds) Communities and Crime Reduction London: HMSO

Burbidge, M. (1981) Priority Estates Projects (1981). Improving Problem Council Estates, Department of the Environment, London: HMSO

Clarke, R.V. and Hough, M. (1984) Crime and Police Effectiveness, Home Office Research Study No. 79, London: HMSO

Coleman, A. (1985) Utopia on Trial, London: Hilary Shipman

Cornish Report (1965) Report of the Committee on the Prevention and Detection of Crime, Home Office, September 1965

Ebklom, P. (1987) Getting the Best Out of Crime Analysis, Crime Prevention Unit Paper, in preparation

Heal, K. and Laycock, C. (1986) (eds) Situational Crime Prevention: from Theory into Practice, London: HMSO

Hill, N. (1986) Prepayment Coin Meters: a Target for Burglary, Crime Prevention Unit Paper, No. 6, available from the Crime Prevention Unit, Home Office

Home Office (1984) Home Office Circular 8/84

Home Office (1985) Home Office Report of the Working Group on Revised Guidelines for Crime Prevention Panels, available from the Crime Prevention Unit, Home Office

Home Office (1986a) Reports of the Standing Conference Working Groups, available from the Crime Prevention Unit, Home Office

Home Office (1986b) Crime Prevention and the Community Programme: a Practical Guide, available from the Crime Prevention Unit, Home Office

Hope, T. (1986) 'Crime, community and environment', Journal of Environmental Psychology, 6, 65-78

Hough, M. and Mayhew, P. (1983) The British Crime Survey: First Report, Home Office Research Study, No. 76, London: HMSO

Houghton, G. and Berry, G. (1987) Microcomputers - an aid to Crime Analysis, available from Home Office Scientific Research and Development Branch, Horseferry House, Dean Ryle Street, London SW1P 2AW

Poyner, B. (1983) Design Against Crime, London: Butterworths

Shapland, J. and Osborn, S. (1987) 'Crime prevention on English housing estates: policies and practices', paper written for SEC Newsletter, April 1987

Smith, L.J.F. and Laycock, G. (1985) Reducing Crime: Developing the Role of Crime Prevention Panels, Crime Prevention Unit Paper, No. 2, Home Office, Crime Prevention Unit

Chapter Sixteen

CRIME PREVENTION: THE NORTH AMERICAN EXPERIENCE

Patricia L. Brantingham

Crime prevention has long been espoused as the principal goal of the North American criminal justice system. It has been the stated first priority of police forces since their inception. It has long been identified by legal authorities as the principal purpose of the criminal law. It has been identified as one of the principal goals of virtually every correctional programme. It has also been consistently advanced as one of the stated goals of such diverse social programmes as slum clearance, compulsory education, vocational training, and mental health screening (Brantingham and Faust 1976, 286-7). In fact, the vast array of social programmes mounted in the American 'war on poverty' in the mid-1960s, for instance, was justified on crime-prevention grounds (Moynihan 1969). Crime prevention is undertaken by all levels of government and by a wide range of community groups. There is no one focus for crime-prevention activity. While the police have been a major force in programming and the federal government the source of most research funds, there is no single crime-prevention policy. While there is no single centre, there is a pattern of activity that could almost be called a crime-prevention movement.

The 1960s and 1970s in North America witnessed both a major expansion of explicit crime-prevention activities and the development of widespread enthusiasm about the ability of individuals and governments to reduce crime levels through these explicit programmes. Over the past twenty years, certain core crime-prevention programmes have been accepted by the general public and have been institutionalized, becoming normal activities routinely pursued by local community groups and by the various agencies of the criminal justice system.

Applied crime-prevention research, studying what works and what does not, began in earnest during the 1970s, at the same time that more basic research on criminal behaviour often adopted a patina of concern for crime prevention. It was widely hoped that a few relatively simple, standardized crime-prevention programmes could, with the right type of research and evaluation work, be devised and developed for general application throughout North America.

In recent years, North American expectations for explicit crime-prevention programmes and research have changed. The belief that a few standardized crime-prevention programmes will work everywhere has been replaced by a more realistic appraisal of the inherent complexity of the criminal event and the realization that generic solutions to local problems probably do not exist. The views of the 1960s and the 1970s are being replaced by an understanding that effective crime-prevention activities must be tailored to a better understanding of both the dynamics of criminal activity and the responses of people and communities to crime. This chapter reviews the maturation of North American thinking about crime prevention over the past twenty years and suggests some directions for crime prevention in the future.

Conceptual models of crime prevention

Since almost any activity undertaken by the criminal justice system can be labelled 'Crime prevention' and since almost any action designed to influence the socio-economic structure to reduce social problems is liable to be characterized as a crime-prevention programme, crime-prevention activities must be understood within some organizing and limiting conceptual framework. During the 1960s and 1970s, substantial effort was put into development of descriptive typologies for analysis and categorization of the different types of activities advanced as crime prevention. More recently, North American crime prevention has been analysed through a process model of criminal activity.

A technique-based model of crime prevention

One approach to describing crime-prevention activities was developed by Peter Lejins in the late 1960s and addressed the techniques employed in various activities. He classified crime-prevention activities as punitive, corrective, or mechanical depending on whether they sought to reduce crime by threats and punishment; or by amelioration of personal, social, or economic conditions thought to impel individuals to commit criminal acts; or by imposing physical barriers making criminal acts much more difficult to commit. His analysis placed most hope on corrective measures and least hope on mechanical crime prevention (Lejins 1967).

The Lejins model continues to provide a useful framework for classification of crime-prevention programmes on the basis of technique, but it was overtaken by developments in the early 1970s. Following the publication of works by criminologist C. Ray Jeffery (1977) and architect Oscar Newman (1972) which emphasized the importance of the environment on the location of crimes and on the decision to offend, much of the emphasis in North American crime-prevention activities came to be placed on the forms of target-specific mechanical and corrective crime prevention. An alternative model for conceptualizing crime-prevention activities on the basis of the target of the activity was advanced by Paul Brantingham and Frederic Faust (1976).

A target-based model of crime prevention

The Brantingham and Faust approach adapted the public health model of disease prevention to the description and analysis of crime-prevention activities. Primary crime-prevention activities are targeted at conditions in the physical and social environment that provide opportunities for or precipitate criminal acts. The objective is to reduce the extent of the precipitating conditions and the frequency of criminal opportunities. Secondary crime-prevention activities are targeted at high-risk settings and high-risk individuals. The objective is to intervene in the setting or in the individual's life so that a crime is never committed. Tertiary crime-prevention activities are targeted at specific locations and offenders after the commission of a crime. The objective is to intervene so that no further offences are committed.

The Brantingham and Faust model served as a useful device for classifying crime-prevention programmes in the 1970s and early 1980s (Hylton 1982). It noted that Lejin's crime-prevention techniques were directed at all three levels of prevention objectives. It helped focus research on the objectives of particular crime-prevention programmes, rather than on the technique employed, and so proved helpful in evaluation research where the emphasis is placed on trying to determine whether objectives are met. A growing body of research on crime prevention, however, suggests that another framework would be even more useful in understanding how North American crime-prevention activities have developed over the past twenty years.

A criminal process model of crime prevention

Crime-prevention activities should be understood as interventions in criminogenic situations that are intended to stop individuals from committing crimes. Activities undertaken in the name of crime prevention may be more easily understood if analysed through a model of the process through which criminal behaviour occurs. It is then possible to explore the points in this model at which intervention may have substantial effects. At its simplest, criminal activity can be thought of as having three phases: a decision phase; a search phase; and a criminal act phase. Each phase provides opportunities for interventions that are likely to have crime-prevention effects (Brantingham and Brantingham 1978; Clarke and Cornish 1985; Taylor and Gottfredson 1986).

In this simple framework, the decision phase is the period during which an individual or a group of individuals decides to commit an offence. This decision may involve conscious, rational deliberation, as appears to be the case in many bank robberies, or it may be essentially emotional, irrational, and instantaneous, as seems to be the case in many assaults (Brantingham and Brantingham 1978, 1984).

The search phase may be short or protracted, but basically involves locating a potential victim in time and space. In an assault this probably means attacking someone a few feet away. In a burglary the process is usually more protracted and may involve several stages: working oneself up to committing the crime; locating a target area; and choosing a specific target within that area (Brantingham and Brantingham 1978; Brown and Altman 1981; Duffala 1976;

Taylor and Gottfredson 1986; Cornish and Clarke 1986; Rengert and Wasilchick 1985; for English studies in accord, see Maguire and Bennett 1982; Bennett and Wright 1984; 30 et seq.). In the case of a crime such as soliciting, the search for a target involves the prostitute choosing a place to wait for potential customers and then choosing between them when they appear (Lowman 1984a, 1984b).

No matter what the crime, there is a criminal act phase. Once a target has been identified, an attempt to commit the crime occurs. Depending on the circumstances, it may or may not be successful. In fact, the attempt itself normally constitutes a crime in its own right under most North American penal codes (see, e.g., Criminal Code of Canada sections 24, 421; California Penal Code sections 663, 664). The criminal act ordinarily occurs very quickly, taking a few seconds in an assault, a few minutes in a burglary. Only rarely does it take an extended period of time to complete.

Intervention in the decision phase

Many kinds of measures undertaken by the state, by private groups, and by individuals can be considered intervention activities that are intended to influence people against the decision to commit a crime. Law-making, with its implicit belief in the deterrent effect of proscribed punishments; social prevention programmes which try to eliminate the 'causes' of crime; and education programmes which try to teach the dangers of criminal activity can all be considered interventions designed to influence the decision to commit an offence.

Law-making

Laws are supposed to deter by defining prohibited behaviours and by specifying the penalties that will be imposed on those who nevertheless engage in them. In North America the potential deterrent effect of law is most frequently mentioned in debates about the death penalty, additional penalties for the use of firearms in the commission of a crime, and life sentences for persons convicted of trafficking in drugs.

Social prevention

The decision to commit a crime is also the principal point of intervention for most social prevention activities. Understanding why people commit crimes has always been a core purpose of criminology. Criminological theory tries to explain why some individuals commit offences and others do not, or why some groups have higher offence rates than others. Offender-centred criminologists search for motives and reasons for criminal behaviour in social conditions, friendship networks, family environments, and parenting patterns; in inherited traits and learned attitudes; in the impact of laws and the behaviour of the justice system and so forth. Offender-centred criminological research looks for correlations between these presumed predisposing conditions and the decision to engage in criminal behaviour (see Herbert 1976; Herbert and Hyde 1984).

Social prevention is generally tied to beliefs about the causes of crime and involves identifying conditions thought to cause crime, then changing those conditions. The quintessential social prevention programme in North America was the Chicago Area Project launched in the 1930s by Clifford Shaw and made famous by the research he and his associates conducted through the Institute for Juvenile Research in Chicago (Shaw and McKay 1929, 1931, 1942, 1969; Shaw and Moore 1931).

Shaw concluded from his early research that the 'treatment' of delinquency in a large city requires changing the conditions in both smaller local communities and whole larger sections of the city. Shaw believed neighbourhood residents had to be organized so that the natural forces of social control could work. In 1932, the Chicago Area Project established twenty-two neighbourhood centres in six areas of Chicago. These centres employed local residents and were locally controlled. The centres had three major functions: to sponsor local recreation and community activities; to co-ordinate local community organizations such as churches, schools, and clubs; and to co-ordinate efforts to improve housing and the physical appearance of the areas (Shaw and McKay 1942, 323-4). While the Chicago Area Project continued until 1957, its impact in reducing crime and delinquency was never precisely evaluated.

In the course of the project vast quantities of information were collected in an attempt to explore Burgess's (1925) zonal hypothesis that, barring intervening features, a city would develop outward from a central business district in a series of concentric zones with distinctive land uses and populations which could be differentiated by socio-economic characteristics. The zonal model predicts that rates of criminal residence will be highest towards the core and will decline with distance from the core. The model also predicts that 'natural areas' of criminal residence should develop within each of the zones (Brantingham and Brantingham 1981, 13).

The work of Shaw and McKay is not known for demonstrating the crime-prevention effects of the local social prevention programmes but rather for its theoretical impact and empirical approach. It formed a bridge between human ecology as conceived by Park and Burgess and those sociologists more narrowly interested in crime. In fact, Harold Finestone (1976, 1977) has argued that virtually all theoretical and empirical work since 1930 can be tied back to reactions against or extensions of the work of Shaw and McKay.

Social prevention activities are not, of course, circumscribed in time, in space, or in technique, by the Chicago Area Project. Social prevention has been supported by such turn-of-the-century moral crusaders as the prohibitionists in both the United States and Canada, and by the more recent liberal crusaders of President Johnson's 'war on poverty' in the 1960s. It has recently been the focus of a major programme undertaken by the Ministry of the Solicitor General in Canada. This ministry has launched a multi-year project that through the use of local groups, tries to identify 'women in conflict with the law' and women who by reason of their socio-economic condition, are likely to come in contact with the law. These local groups have subsequently been given funding in order to enable them to design and develop projects specifically aimed at helping those women who have been identified during the first phase of the programme.

As the previous examples indicate, social prevention in North America frequently takes on a local face and is initiated by community and church groups. As a result, these social prevention programmes are too numerous to count. A proto-typical example is a series of social prevention programmes undertaken by the West Island YMCA in Montreal, Quebec. In the mid 1970s this organization set up a series of community and volunteer-based programmes designed for 'hard to reach' youth. The programmes include supervised community work, group discussions with school leavers that were intended to stimulate interest in education and educational opportunities; courses on 'life-skills'; and - probably simply to attract youth - courses on motorcycle riding (Brantingham 1986).

Social prevention programmes are also frequently aimed at youth who have already come into conflict with the law. Such programmes are often directly run or subsidized by governments and range from counselling and other treatment approaches for non-incarcerated offenders to socially-oriented diversion programmes and programmes for problem families. Research into the impact of social-prevention programmes is difficult, since programmes are numerous, frequently local in nature and highly varied. The variability coupled with the small size of most programmes, makes research almost impossible. The individual programmes are so highly varied that even if most of them had been completely evaluated, it would probably be impossible to say what might or might not generally work. As might be expected, those few competent studies that do exist tend to show mixed results: some programmes in some locales seem to work and some do not (Hiew 1981; Hackler 1978; LeBlanc and Frechette 1986; Finckenauer 1982; Lundman 1984).

Education

The goal of changing motivation and thereby affecting the decision to commit an offence is behind many North American advertising and public-education programmes. Massive media campaigns against driving when drunk, against the use of drugs, against shoplifting and against vandalism are popular all across North America, although research provides weak support at best for the efficacy of such approaches (Sacco and Silverman 1981; Sacco 1985; for an English study see Riley and Mayhew 1980). Educational programmes aimed at altering the decision to offend range from formal courses on the law and the consequences of a criminal record taught in the schools to occasional public lectures by the police, by lawyers, or by judges. As with media approaches, there is little evidence that such programmes reduce the incidence of criminal behaviour, but as with social prevention programmes they are popular perhaps because they offer the hope of intuitively appealing, simple solutions to complex problems (Albert and Simpson 1985; Griffiths 1982; Hiew 1981).

Analysis of programmes aimed at the decision phase

North American programmes aimed at altering the decision to commit an offence range across the categories of primary, secondary and tertiary crime prevention (Brantingham and Faust 1976). Some approaches, such as general education programmes, are aimed at large cross-sections of the population. It is hoped that within that cross-section there will be some individuals who will receive the message and alter their behaviour. These programmes involve primary prevention since the education is given to broad cross-sections of the population. Other programmes, such as the 'women in conflict with the law' programme in the Ministry of the Solicitor General or programmes similar to the ones at the West Island YMCA in Montreal can be classified as secondary prevention programmes. These are aimed at 'high-risk' groups, that is, groups with higher than average rates of offending. Prevention programmes designed to alter the behaviour of individuals who have already committed an offence such as juvenile diversion programmes or prison treatment programmes are tertiary activities. As one moves from primary through secondary to tertiary prevention activities, the target of the activities narrows from the general public or broad cross-sections of the population to specific individuals thought to have a high risk of offending.

Programmes designed to prevent or reduce the amount of crime by altering the motivation behind criminal acts or by altering the decision to commit a crime have had limited success and will continue to have limited success until we know more about what influences the decision to commit a crime. Thanks to Shaw and McKay and their successors in empirical criminology, we have amassed a substantial amount of knowledge about the correlates of criminal behaviour in North America and we learned much about the characteristics of concentrations of residences of criminals. Not much is known, however, about the actual decision to offend (see Cusson, 1983, for an approach which does address the decision to offend). Consequently, most crime-prevention activities that are intended to intervene in the decision to offend are not based on knowledge of what influences that decision, but rather try to alter various underlying conditions that are commonly correlated with criminal behaviour and are assumed to be causally linked to the decision to commit a crime. At best such programmes use very simple models of decision behaviour. Various programmes try to reduce youth unemployment, to provide youth with alternative activities that are law-abiding, to provide supplementary education or to provide information about the consequences of criminal behaviour in the implicit belief that such actions will change the contingencies that shape the decision to offend. More research is needed on how people do decide to commit a crime.

A recent model of offender decision-making has been presented by Clarke and Cornish (1985). Their model is specified for residential burglary. In their model they argue that the decision to offend first involves an individual's recognition of his 'readiness' to commit an offence to satisfy his needs for money or goods or excitement. This readiness is influenced by background factors such as age, sex, and race, as well as predisposing factors related to upbringing and previous experiences. The actual decision to commit a burglary may be influenced by some chance event such as the need for money at a specific point in time, pressure from friends who suggest a burglary (Duffala 1976) or seeing a target in the course of usually law-abiding activities (Letkemann 1973; Bennett and Wright 1984; Maguire and Bennett 1982).

Cornish and Clarke argue that the factors influencing the decision to commit a crime vary with the detailed characteristics of the crime. If this is the case, and it seems very likely it is, prevention activities designed to influence the decision to commit a crime would have to be highly specialized and particularized for each type of crime and for a variety of interacting factors influencing 'readiness' as well as precipitating factors. It would be unlikely that primary social prevention or primary educational programmes would have much measurable effect. Programmes would have to be targeted for both potential offenders and precipitating factors.

Targeting would, at least logically, increase the probability of a particular crime-prevention activity having some influence on the decision to commit a crime. Narrow targeting, however, reduces the likelihood that any one prevention strategy will have a broad-based effect. This is particularly true for prevention strategies targeted at convicted offenders. Such a low proportion of crimes actually end in an arrest and conviction in Canada and the United States that even the most successful offender-oriented prevention programme would be unlikely to have much impact on aggregate crime patterns.

Intervention in the search phase

Once the decision to commit an offence has been made, or perhaps concurrent with that decision, the potential offender must locate a target or a victim. This target search and identification process has been better researched in North America than has the decision to commit a crime. Intervention in this search process is the goal of the standard core crime-prevention programmes most often undertaken by the police and by community groups.

Core crime-prevention activities

Over the last fifteen years several programmes have been widely accepted as standard off-the-shelf crime-prevention activities. The most widely adopted of these core crime-prevention programmes depend on some type of community (i.e. non-police) surveillance to provide security for buildings and protection against property crime. The best known of these programmes is called neighbourhood watch or block watch. In fact neighbourhood watch has become synonymous with crime prevention in the minds of many. Neighbourhood watch programmes, which have now been adopted in many countries, involve organizing small residential areas, sometimes residential units as small as a single apartment building, into local committees. These committees in turn attempt to convince the residents that they should become the watchers for the neighbourhood and should call the police if they see anything amiss. Areas with neighbourhood watch programmes frequently have large signs at access points warning people that: 'This neighbourhood is protected by neighbourhood watch'.

The neighbourhood watch approach usually involves some type of property marking (called operation identification in most parts of North America) and the placing of decals on windows or doors to inform the would-be thief that goods are marked and therefore identifiable. Some neighbourhood watch programmes seem to work; others do not.

Neighbourhood watch has triggered several similar programmes: marine watch, where owners of boats are asked to watch for signs of trouble in and around the marinas where they are berthed; realtor watch, in which real-estate agents watch out for signs of criminal activity at vacant houses; and several other similar programmes that depend on persons such as taxi drivers, bus drivers, and telephone repairmen, who routinely travel through residential areas keeping a look out for suspicious activity.

Neighbourhood watch, operation identification and related programmes attempt to disrupt the intending criminal's search for targets. It rests on several assumptions. First, that police patrol is unable to provide adequate levels of surveillance and that residents must replace or supplement official watchers such as the police. Second, that intending criminals who are looking for a target and know that a particular area is supposed to have higher than normal levels of surveillance will avoid that area because the potential risk is too high. If this second assumption is correct, criminals, as the result of being put off from one target, may be displaced to some other less secure or less thoroughly watched target. The third assumption is that marked goods are both more difficult to dispose of and easier to identify and use as evidence, and will therefore be avoided by criminals. Tied to this assumption is the further assumption that intending criminals will make rational choices and avoid both whole neighbourhoods with neighbourhood watch programmes and individual houses displaying decals warning that goods are marked. Underlying all these assumptions is the belief that communities can defend themselves.

North American research on the efficacy of neighbourhood watch and related approaches has been mixed, with the balance possibly shifting slightly towards the positive (Perry 1984; Worrell 1984; Hackler 1978; Normandeau and Hasenpusch 1978; Titus 1982; White et al. 1975). It seems to work in some neighbourhoods and not in others. It seems easiest to organize in middle-class neighbourhoods where crime rates are generally low and seems to be most difficult to organize in poorer neighbourhoods which frequently have higher crime rates.

Another core community-based programme involves citizen patrols. Such patrols are more popular in the United States than in Canada. Patrols are often set up after a high-profile criminal incident in a residential area. They usually involve limited numbers of individuals who patrol on foot at night or during weekends (Kelling 1986; Einstadter 1984). The persons on patrol are not supposed to be armed or to attempt to stop crimes in progress. Patrols seem to be a response to perceived problems with crimes against persons such as robbery and assault rather than a response to property crimes such as burglary. Patrols are obviously based on a belief in the deterrent effect of active and obvious surveillance and the fear-reducing potential of the presence of legitimate watchers.

Persons on patrol may be untrained or may be an adjunct to regular police (Sherman 1986; Marx and Archer 1971; Klockars 1985). For example, New York City had, in 1986, almost 9,000 volunteers who were certified by the police, armed with night sticks and used on regular patrol. New York city is also the centre of the Guardian Angel movement. Guardian Angels patrol public places and subways while wearing para-military dress (Kenney 1986). It is perhaps ironic to reflect that modern police had their origins in semi-official patrols such as the Bow Street Runners which were organized to deal with problems of street robbery and assault in late eighteenth-century cities.

Both neighbourhood watch and community patrols are frequently organized by the police. When this happens the underlying crime control/deterrence rationale comes from the police. Community members become extended eyes and ears of the police. Their presence is believed either to deter criminals or to aid in their apprehension. Most North Americans seem to believe that crime comes from outside their own neighbourhoods. As a consequence, these prevention programmes are aimed at outsiders. For example, many areas with neighbourhood watch programmes place large bulletin boards announcing the existence of the programme at major access points to the neighbourhood. These signs are designed to warn criminals away.

Sometimes these programmes are brought into existence fairly independently of the police as part of the community crime-prevention movement. The community crime-prevention movement appears to be built on the belief that 'communities', however they are defined, either contain the elements that protect them from crime or lack those elements and are consequently high crime communities. Community action usually involves organizing residents in some loose fashion, holding information sessions, occasionally engaging in fund-raising activities and engaging in some type of crime-prevention activity.

Under these types of programmes communities are supposed to be strengthened from within and, with this increased strength, resist forces producing or permitting crime from within. This approach is consistent with a long tradition in criminology linking the importance of the structure and characteristics of a community to its crime levels. This tradition is based on the well-established empirical fact that crime rates vary by area and that the variation is correlated with the social, demographic, and organizational characteristics of the communities that exist within different areas.

The United States Department of Justice has been very interested in evaluating community crime-prevention efforts. Two major crime-prevention programmes involving non-governmental organizations have been funded through the Law Enforcement Assistance Administration beginning in 1977. One programme, the Comprehensive Crime Prevention Programme (CCPP), involves the co-ordination of public and private crime-prevention activities in seven of the largest cities in the United States. The other programme, also started in 1977, was called the Community Anti-Crime Programme (CAC) and involved the funding of 146 local community crime-prevention groups at a cost of $30 million in the first year alone. Studies of these projects and others like them find mixed benefits at best. The most frequently demonstrated benefits take the form of reduction in the levels of fear of crime rather than reduction in reported crime rates (Washnis 1977; Kenney 1986).

Planning and architectural approaches

People express many needs in the cities and buildings they build. Our cities are a window on our culture. They are also a window on the problems of a society. A look at cities throughout history reveals a persistent concern about security. In different historical periods cities have been built within walls for protection; in almost all cultures individual houses or buildings have been fortified in one way or another. People build defensively, though the degree of defensiveness varies. The intent is to intervene in the target searches of intending criminals by imposing access barriers and signalling vigilant defensive postures.

Town planning has also been affected by the desire to intervene in the criminal search process. The redesign and reconstruction of Victorian London was, in substantial measure, shaped by the desire to break up the 'rookeries' - neighbourhoods of dense criminal residence strategically perched on the city's edge with easy access to a variety of rich targets (Tobias 1972). Zoning of land uses and minimum configuration of apartments in many North American cities also were undertaken, in part, to eliminate conditions and environmental factors believed to be causes of crime.

A conscious tie between crime and architecture or town planning weakened in North America after the Second World War with the boom in building of suburbs and the advent of commercial strip development. Accessibility and openness became quite important. While there is regional variation across North America, most residential areas were built with common areas and few walls or fences around houses. Meandering green spaces were frequently included in housing developments, increasing the general accessibility of areas. The standard suburban houses and suburban apartments were built with highly accessible windows and sliding glass patio doors. While houses and buildings still had locks on doors and windows, the main emphasis in building was on openness.

Security was not a major concern in architecture or planning. This lack of interest in residential security and in the relationship between architecture, town and country planning, and crime began to change in the 1970s, under pressure from a rapidly increasing crime rate, when three books were published. One by the planner-architect Oscar Newman entitled Defensible Space: Crime Prevention Through Urban Planning, another, Crime Prevention Through Environmental Design, by C. Ray Jeffery (1977; Jeffery et al., 1985) and the third by Jane Jacobs entitled The Death and Life of Great American Cities (1961) brought the relationship between the physical environment and crime occurrence into the general debate on crime prevention.

Newman's first book (he wrote several others) and the films based on this book have perhaps had the greatest impact. They seemed to offer a simple, straightforward method for reducing crime, particularly for reducing crime in the deteriorated areas of inner cities. His ideas were very appealing. In essence, he argued that the space we live in, that is, the area immediately around our homes and our near-neighbourhood, can be made safer and more secure by designing it so that it can be controlled by the residents. He argued that the crime in an area was directly related to the quantity and quality of the uncontrolled public space in and around residential buildings. He argued that by dividing the space in and around residential buildings into small areas which are functionally under the control of individual residents a type of local social control would be induced and, as a consequence, crime would be reduced.

Newman's early works have been roundly attacked on methodological and theoretical grounds (Mawby 1977; Mayhew 1979; Taylor et al. 1980), but despite these attacks, his work and that of Jeffery and of Jacobs proved to be a catalyst. Since the early 1970s, criminologists, geographers, and planners have become interested in the theoretical relationships between the environment and the patterns of crime.

North American research on the relationship between environment and crime has tended to follow a model somewhat similar to, but more fully elaborated than, the model used in this chapter. Research has centred on exploring the choice and search behaviour of criminals and on examining the capacity of residential areas or individual buildings to give off cues to a potential criminal that tell the criminal to stay away. This research has involved both the development of theoretical models and conduct of empirical studies of the spatial behaviour of criminals - how they choose an operating area and a specific target within that area - and the converse development of models and conduct of empirical studies that try to determine what it is about a particular area or individual target that makes it attractive to intending offenders. The research has basically found:

(i) That the pattern of target choice is complex, but in the aggregate probably can be modelled empirically
(ii) That the search space for many criminals is limited and probably limited to areas in and around their normal, law-abiding activity areas
(iii) That property criminals probably identify an area first and then search or wait for a target in that area
(iv) That the actual choice of an operating area and a specific target involves the assessment of many patterned cues.

(For a general review of this literature see Brantingham and Brantingham 1984).

Governments, both at national and municipal levels, have become interested in using architectural and planning techniques to reduce crime. Both the Department of Justice in the United States and the Ministry of the Solicitor

General in Canada have funded research and demonstration projects to try to determine if clear, strong relationships between architectural or planning changes and crime occurrence can be found. In fact, efforts have been made to develop crime-preventive architectural design guidelines (see, for example, Newman 1975, 1980; McInnis et al. 1982; Gardiner 1978). Because of the variety of the socio-geographic factors influencing target choice, the guideline approach has not proved very fruitful.

Probably the best known design-oriented project was conducted in Hartford, Connecticut, where changes were made to the street network in a high-crime area. Other changes were also made. Most notably community development projects were introduced and the police became more actively involved in the area. Initial evaluation of this project showed a substantial decrease in areal crime rates (burglary decreasing 42 per cent and robbery decreasing 27.5 per cent). Subsequent evaluation showed less-enduring impact (Fowler 1981; Fowler and Mangione 1982).

While other projects aimed at manipulating the physical environment produced mixed results, research continues to show examples of the strength of the relationship between the environment and crime. (For recent North American research see Stoks 1982; Greenberg et al. 1984; Rengert and Wasilchick 1985; Weaver and Carroll 1985; Beavon 1984; Wilson and Kelling 1982; and in particular the research on armed robbery undertaken at the International Centre of Comparative Criminology at the University of Montreal). The relationship, however, is perhaps as complex as the relationship between the social environment and crime. In fact, as Newman later came to argue (Newman and Frank 1980), the physical environment by itself cannot be used to explain the volume or perhaps even the distribution of crime. It is the environment as a whole including the social, physical, cultural, legal, and governmental aspects which must be considered when looking at crime. Interesting and important research can be undertaken by exploring one aspect of the environment and 'holding the other aspects constant', but effective crime-prevention research needs to consider all aspects.

One architectural-planning approach undertaken by the Royal Canadian Mounted Police (RCMP) in the province of British Columbia is worth noting. In Canada, the RCMP is a federal police force, but it also provides municipal policing services under contract. In British Columbia there are many RCMP detachments providing municipal services. The senior RCMP officers in British Columbia support the use of architectural and planning approaches to reduce crime and 'incivilities'. They do this in an interesting way. RCMP crime-prevention constables are trained in techniques of crime analysis and given courses on environmental criminology and town planning. They are taught an approach similar to the situational crime-prevention approach of the Home Office in England as exemplified in the approach used by Barry Poyner and described in his book Design against Crime (1983). After the courses, they return to their detachments and are encouraged to become involved in the local planning process, especially reviews of applications for development permits. They carry out site-specific reviews before buildings or developments are put up or before planning decisions are made. They also help municipal officials find solutions to trouble spots. (For an example of this approach see Hest and Harrison 1983).

The successes reported by these crime-prevention officers tend to be highly site-specific. Proposed solutions depend on the detailed characteristics of the problem. The approach is really a process approach and work frequently involves modifying the plans for proposed developments. As a result, evaluation of the effectiveness of any particular change is difficult. Moreover, there is always the possibility the problems will just be displaced.

Architecture and the law

There is another movement in North America which will probably prove to be a major force in crime prevention in the near future. Individual victims of crime have begun to sue developers and landlords successfully on the grounds that built environments have proved to be unsafe. While the successful cases are rare in Canada, they are becoming more frequent in the United States.

Most court actions are based on the doctrine of 'Warranty of habitability'. Jeffery and White (1985) provides a review of cases in the United States (see also Muth 1984). Jeffery cites cases including ones involving innkeepers and universities being held liable for assaults on hotel guests. One example, Walkoviak v. Hilton Hotels 580 S.W. 2nd 623 (1979), is instructive. In this case Hilton Hotels was held liable for an assault on a guest who had to walk out of the hotel to reach his car in a poorly lit parking lot. The court held that the hotel chain should have been aware of the danger of the situation.

While the number of cases is still limited and no good parameters have yet developed as to what level of protection should be provided, and while courts are not yet uniformly accepting such arguments, these types of suits will almost certainly increase in the future and as they do, they will almost certainly modify North American building patterns.

Analysis of programmes aimed at the search phase

Crime-prevention programmes aimed at altering the target search behaviour of criminals may offer the most hope of success in reducing crime. Unlike programmes aimed at altering the decision phase, there is both a better-developed theoretical structure and more supporting research (Brantingham and Brantingham 1981, 1984; Brown and Altman 1981; Taylor and Gottfredson 1986). Researchers are beginning to understand actual search behaviour. Research in geography, planning, and architecture shows that spatial behaviour can be altered. There is much less evidence that programmes can be developed which successfully alter the decision to offend.

The major problem with programmes designed to alter target search behaviour involves both the locational specificity of many programmes and underlying problems with state involvement. For many types of crimes, search behaviour involves many stages (Brantingham and Brantingham 1978; Brown and Altman 1981). At each stage there is a narrowing of potential targets until one is picked. Crime-prevention programmes can be designed to intervene at any stage in this process. The more narrow the focus, however, the more likely the criminal is to pick another target or to come back to the same target or operational area at some other time. It is possible to protect an individual house from a burglary, but protecting that house may increase the risk to a neighbour's house. Unit-specific protection, be it a house, a car or a store, may have no effect on the overall crime rate. There is the potential for displacing criminal activity.

Research on displacement is conceptually very difficult and, to date, limited in North America (Lowman 1983;

Gabor 1978). Models of search behaviour should at least consider the potential for displacement. Models for understanding crime displacement must be elaborated and eventually must be integrated into models of target search behaviour. It seems, almost without the need of support, that crimes are most displaceable if: (i) there are other attractive targets easily accessible (either nearby or available at another time of the day); or (ii) if the potential criminal is strongly motivated. It is probably true that most crime-prevention programmes designed to influence the search phase produce both some displacement and some abatement of criminal activity.

Because of the potential for displacement, there is a serious question of the responsibility of the state when it engages in crime-prevention programmes. If crime is merely displaced from one target to another by a prevention programme, then there is no aggregate gain. More seriously, if state action protects some but increases the risk for others, the state should consider whether the modification of relative risks for different groups is acceptable. For example, if government supports crime-prevention programmes to protect store owners from burglary, but the burglars just move to nearby houses, serious questions have to be asked about the benefits of the programme. Much more research and evaluation is needed to understand how crime-prevention programmes which attempt to alter search behaviour change the relative risk to different groups of people (Brantingham and Brantingham 1984). Understanding displacement and its policy implications is an area in need of a great deal of research.

The criminal act phase

The final phase in the model presented in this chapter is really the last step in the search phase. It has been identified as a separate phase to emphasize a series of unique crime-prevention approaches. Interventions designed to interfere with the criminal act generally involve target-hardening strategies. The criminal act itself is made more difficult physically. Locks, bars, chains, walls, fences, automobile steering-wheel locks, are obvious attempts at target hardening.

Target hardening is used to protect individual targets. Displacement potential is, therefore, high unless most targets are protected. The experience with steering-wheel locks on automobiles is perhaps the best documented case of the usefulness of a target hardening technique when widely applied. In Germany where all automobiles, both old and new, were fitted with steering-wheel locks, there was a real and sustained decrease in automobile thefts. In Great Britain where only new automobiles were fitted, there was a decrease in theft of new vehicles and increased thefts of older vehicles (Clarke and Mayhew 1980).

Target hardening is always possible, but may not always be desirable. Target hardening, from a societal perspective, seems to be less desirable if it involves increasing societal levels of fear of crime. Target hardening in residences frequently involves individuals installing additional locks and bars or installing a burglar alarm system. Since increasing people's fear levels may be a necessary precursor to selling target hardening in many instances, its value must be judged against actual crime risk levels. High fear levels can have negative social effects by making people afraid to leave their homes, reducing social interaction levels, or by persuading individuals to spend money unnecessarily.

Canada and the United States differ in their use of fear as a motivation for crime prevention. This is most notable in the publicity that surrounds neighbourhood watch. In the United States the most common symbol for neighbourhood watch is an outline of a menacing figure in cloak and hat. In Canada the symbol is a profile of a person blending with a non-threatening representation of a police officer together with the slogan 'working together to prevent crime'. The difference is perhaps most accentuated in British Columbia where the RCMP have taken the policy position that they will work through crime-prevention programmes which support (in their words) an 'urban village' approach, not a 'fortress mentality'.

New advances in target hardening

There are continual advances in the technology of target hardening. For example, target hardening has become a major concern in the computer field, especially in telecommunication. Call-back devices have been developed to minimize illegal entry into computer networks; hard wired lines are used instead of dial-up systems; and computers are even placed in lead-lined rooms to prevent electronic eavesdropping.

Housing security advances are always being made. The latest move in North America is the general marketing of alarm systems to the middle class. As an example, in Ottawa, Canada, an urban area of about 800,000 people, there are sixty-nine separate burglar alarm companies providing services to the public. At the same time that home burglar systems are becoming more popular, basic housing design in North America remains open and accessible.

The other general trend in North America is towards increased use of anti-shoplifting devices: exploding tags; alarm systems tied to tags on goods; viewing devices in changing rooms; chains through sleeves of coats; and increased store security, both through cameras and individual watchers.

Analysis of programmes aimed at the criminal act phase

Programmes aimed at interfering with specific criminal acts are most likely to work in protecting individual targets, but are unlikely to have any aggregate impact on the level of crime unless widely adopted. These are the types of approaches most likely to produce geographic, temporal, or crime-type displacement. As a result, use of target-protection techniques will have an impact on aggregate crime levels only if part of a wider programme.

Target-protection programmes highlight the importance of distinguishing between individual crime-protection strategies and crime-prevention strategies aimed at protecting people in general. These approaches, because of the potential for displacement, bring into focus the potential conflict between individual safety and group security. It becomes particularly problematic because individual crime-prevention strategies are expensive and therefore generally only available to members of the higher economic strata of society.

Conclusion

Crime prevention seems to be proceeding at two levels. Standardized, off-the-shelf crime-prevention programmes continue to be popular whether they be community-level social programmes, police-sponsored surveillance schemes or target-hardening approaches. At the same time that programming is becoming more standardized, theoretical and applied research in the North American context indicates that crime-prevention strategies must be made highly specific to a given socio-geographic environment if they are to have any appreciable effects. Criminal behaviour is complex and varied. Research indicates that standardized programming is unlikely to work as a general approach to preventing crime.

It seems fairly clear that more research on criminal choice behaviour is needed. Standard crime-prevention programmes sometimes work but, more often than not, do not work. More and better research into the complex of environmental factors influencing crime - social, economic, geographic, physical, or legal - is needed to determine why some approaches work and others do not. Experience over the last twenty years in North America indicates that research focused at the search phase is most likely to produce crime-prevention approaches that reduce crime levels.

There is, of course, a 'catch-22'. As research improves and is driven by better models of criminal choice behaviour, the inherent uniqueness of criminal events begins to surface. For example day-time thefts from the gardens of houses along a pathway taken by teenagers from school to a local after-school hangout are different from thefts from vehicles in a parking garage under a shopping centre. Thefts in the first example may be very opportunistic, involve an obvious and mutable search pattern. In the second example, the persons breaking into automobiles are responding in a less opportunistic way and because of that their search pattern may be different and less amenable to change. A crime-prevention strategy that addresses the first problem would be different from one for the second problem. As strategies become more unique it is more likely they will work, but it is also more difficult to determine their effectiveness. As the scale reduces, the number of crimes being manipulated also reduces and research becomes more difficult. Research becomes even more difficult if an approach similar to the one used by the RCMP in British Columbia is used. When design changes are made before a development is built, as is the case in British Columbia, it is impossible to determine what the problem would have been had there been no change. The effectiveness of the change cannot be determined directly.

If research continues as it has in North America, some new crime-prevention approaches taking into account what is learned about the patterns of criminal activity will probably be developed in the next five to ten years. Most likely these will be based on a better understanding of criminal choice behaviour.

While changes may occur, they most likely will be additions to the range of crime-prevention programmes now available. The core community and police-based programmes will likely remain popular. Social programming with a crime-prevention focus is not new; surveillance programmes similarly have a long history. With the range of community- and government-sponsored groups engaging in social programming or some type of surveillance activity these approaches are not likely to disappear.

References

Albert, W.G. and Simpson, R.I. (1985) 'Evaluating an educational program for the prevention of impaired driving among grade II students', Journal of Drug Issues 15 (1), 51-71

Bennett, T. and Wright, R. (1984) Burglars on Burglary: Prevention and the Offender, Aldershot: Gower

Beavon, D.J.K. (1984) Crime and the Environmental Opportunity Structure: the Influence of Street Networks on the Patterning of Property Offences, MA thesis (Criminology), British Columbia: Simon Fraser University

Brantingham, P.L., and Brantingham, P.J. (1984) Patterns in Crime, New York: Macmillan

Brantingham, P.L. (1986) 'Trends in Canadian crime prevention', in Heal, K. and Laycock, G. (eds), Situational Crime Prevention: From Theory to Practice, London: HMSO, 103-12

Brantingham, P.L., and Brantingham, P.J. (1981) 'Notes on the geometry of crime', in Brantingham, P.J. and Brantingham, P.L. (eds), Environmental Criminology, Beverly Hills: Sage, 27-54

Brantingham, P.L. and Brantingham, P.J., (1984) 'Burglar mobility and crime prevention planning', in Clarke, R. and Hope, T. (eds), Coping with Burglary, Boston: Kluwer-Nijhoff, 77-95

Brantingham, P.J. and Brantingham, P.L. (1978) 'A

theoretical model of crime site selection', in Krohn, M.D. and Akers, R.L. (eds), Crime, Law and Sanctions, Beverly Hills: Sage, 105-18

Brantingham, P.J. and Faust, F.L. (1976) 'A conceptual model of crime prevention', Crime and Delinquency 22, 284-96

Brown, B.B. and Altman, I. (1981) 'Territoriality and residential crime: a conceptual framework', in Brantingham, P.L. and Brantingham, P.J. (eds), Environmental Criminology, Beverly Hills: Sage, 55-76

Burgess, E.W. (1925) 'The growth of the city', in Park, R.E., Burgess, E.W., and McKenzie, R.D. (eds) The City, Chicago: University of Chicago Press, 47-62

Burgess, E., Lohman, J. and Shaw, C. (1937) 'The Chicago area project', National Probation Association Yearbook, 8

California (1969) Penal Code, sections 663, 664

Canada (1970) Criminal Code, sections 24, 412

Clarke, R.V. and Cornish, D.B. (1985) 'Modeling offenders' decisions: a framework for research and policy', in Tonry, M. and Morris, N., Crime and Justice: An Annual Review of Research 6, 147-85, Chicago: University of Chicago Press

Clarke, R. and Hope, T. (1984) Coping with Burglary, Boston: Kluwer-Nijhoff Publishing

Clarke, R. and Mayhew, P. (1980) Designing Out Crime, London: HMSO

Cornish, D.B. and Clarke, R.V.G. (eds) (1986) The Reasoning Criminal: Rational Choice Perspectives on Offending, New York: Springer-Verlag

Cusson, M. (1983) Why Delinquency? Toronto: University of Toronto Press

Duffala, D.C. (1976) 'Convenience stores, armed robbery, and physical environmental features', American Behavioral Scientist 20, 227-46

Einstadter, W.J. (1984) 'Citizen patrols: prevention or control', Crime and Social Justice 21, 200-12

Finckenauer, J.O. (1982) Scared Straight and The Panacea Phenomenon, Englewood Cliffs, New Jersey: Prentice-Hall

Finestone, H. (1976) 'The delinquent and society: the Shaw and McKay tradition', in Short, J.F. (ed.), Delinquency, Crime, and Society, Chicago: University of Chicago Press, 23-49

Finestone, H. (1977) Victims of Change: Juvenile

Delinquency in American Society, Westport, Connecticut: Greenwood Press
Fowler, F.G. (1981) 'Evaluating a complex crime control experience', in L. Bickman, Applied Social Psychology Annual, Hillsdale, New Jersey: Erlbaum
Fowler, F.J. and Mangione, T.W. (1982) Neighborhood Crime, Fear and Social Control; a Second Look at the Hartford Program, Washington DC: Center for Survey Research
Gabor, T. (1978) 'Crime displacement: the literature and strategies for its investigation', Crime and Justice, 6, 100-7
Gardiner, R.A. (1978) Design for Safe Neighborhoods, Washington DC: US Government Printing Office
Gouvernement du Quebec (1980) Le Vol a Main Armee au Quebec, Quebec: Ministere de la Justice, Gouvernement du Quebec
Greenberg, S.W., Roche, W.M., and Williams, J.R. (1984) Informed Citizen Action and Crime Prevention at the Neighborhood Level, Vol. I-IV, North Carolina: Research Triangle Institute, Research Triangle Park
Griffiths, C.T. (1982) 'Police school programs: the realities of the remedy', Canadian Journal of Criminology 24, 329-40
Hackler, J. (1978) The Prevention of Youthful Crime: the Great Stumble Forward, Toronto: Methuen
Herbert, D.T., and Hyde, S.W. (1984) Residential Crime and the Urban Environment, a report for the Economic and Social Research Council
Herbert, D.T. (1976) 'The study of delinquency areas: a geographical approach', Transactions, Institute of British Geographers, NS1, 472-92
Hest, J.J. and Harrison, J.T.J. (1983) Policing the Port Alberni Harbourfront Project, Port Alberni, British Columbia: Crime Prevention Unit, Royal Canadian Mounted Police
Hiew, C.C. (1981) 'Prevention of shoplifting: a community action approach', Canadian Journal of Criminology 23, 57-68
Hylton, J.H. (1982) 'The native offender in Saskatchewan: some implications for crime prevention programming', Canadian Journal of Criminology 24, 121-32
Jacobs, J. (1961) The Death and Life of Great American Cities, New York: Random House
Jeffery, C.R., del Carmen, R.V., and White, J.D. (1985)

Attacks on the Insanity Defense, Springfield, Illinois: Charles C. Thomas

Jeffery, C.R. (1977) Crime Prevention Through Environmental Design, (rev. ed.), Beverly Hills: Sage

Jeffery, C.R., and White, J.D. (1985) 'Crime prevention and legal liability', in Jeffery, C.R., del Carmen, R.V., and White, J.D. (eds), Attacks on the Insanity Defense, Springfield, Illinois: Charles C. Thomas, 207-16

Kelling, G. (1986) 'Neighborhood crime control and the police: a view of the American experience', in K. Heal and G. Laycock (eds), Situational Crime Prevention: From Theory to Practice, London: HMSO, 91-102

Kenney, D.J. (1986) 'Crime on the subways: measuring the effectiveness of the Guardian Angels', Justice Quarterly 3, 481-96

Klockars, C. (1985) The Idea of Police, Beverly Hills: Sage

LeBlanc, M. and Frechette, M. (1986) 'La prevention de la delinquance des mineurs: une approche integree et differentielle, Annales de Vaucresson 24(1), 87-99

Lejins, P. (1967) 'The field of prevention', in Amos, W. and Wellford, C. (eds), Delinquency Prevention: Theory and Practice, Englewood Cliffs, New Jersey: Prentice-Hall, 1-12

Letkemann, P. (1973) Crime as Work, Englewood Cliffs, New Jersey: Prentice-Hall

Lowman, J. (1983) 'Geography, crime and social control', Vancouver: unpublished Ph.D. thesis, Department of Geography, University of British Columbia

Lowman, J. (1984a) Working Papers on Pornography and Prostitution: Vancouver Field Study of Prostitution, vol. I, report 8, Ottawa, Ontario: Policy, Programs and Research Branch, Department of Justice

Lowman, J. (1984b) Appendices to Research Notes: Vancouver Field Study of Prostitution, vol. 2, WPPP, no. 8

Lundman, R.J. (1984) Prevention and Control of Juvenile Delinquency, Oxford: Oxford University Press

McInnis, P., Burgess, G., Hann, R., and Axon, L. (1982) The Environmental Design and Management Approach to Crime Prevention in Residential Environments, Ottawa, Ontario: Ministry of the Solicitor General

Maguire, M. and Bennett, T. (1982) Burglary in a Dwelling: The Offence, the Offender, and the Victim, London: Heinemann

Marx, G. and Archer, D. (1971) 'Citizen involvement in the

law enforcement process: the case of community patrols', American Behavioral Scientist, 15,52-72

Moynihan, D.P. (1969) Maximum Feasible Misunderstanding, New York: Free Press

Muth, A. (1984) 'Landlord liability for criminal attacks', Trial, March, 72-6

Mayhew, P. (1979) 'Defensible space: the current status of crime-prevention theory', Howard Journal 18, 150-9

Mawby, R.I. (1977) 'Defensible space: a theoretical and empirical appraisal', Urban Studies 14, 169-79

Newman, O. (1972) Defensible Space: Crime Prevention through Urban Planning, New York: Macmillan

Newman, O. (1975) Design Guidelines for Creating Defensible Space, Washington, DC: National Institute of Law Enforcement and Criminal Justice

Newman, O. (1980) Community of Interest, Garden City, New York: Anchor Press

Newman, O. and Frank, K. (1980) Factors Influencing Crime and Instability in Urban Housing Developments, Washington, DC: National Institute of Justice

Normandeau, A. and Hasenpusch, B. (1978) Review of Active Crime Prevention Methods, vol. 2, Montreal: Ecole de Criminologie, Universite de Montreal

Park, R.E. and Burgess, E.W. (1921) Introduction to the Science of Sociology, Chicago: University of Chicago Press

Perry, K. (1984) 'Measuring the effectiveness of neighbourhood crime watch', Law and Order 32 (12), 37-40

Poyner, B. (1983) Design against Crime: Beyond defensible space, London: Butterworths

Rengert, G. and Wasilchick, J. (1985) Suburban Crime: A Time and Place for Everything, Springfield, Illinois: Charles C. Thomas

Riley, D. and Mayhew, P. (1980) Crime Prevention Publicity: An Assessment, Home Office Research Study, No. 63, London: HMSO

Sacco, V.F. (1985) 'Shoplifting prevention: the role of communication-based strategies', Canadian Journal of Criminology 27(1), 15-29

Sacco, V. and Silverman, R. (1981) 'Selling crime prevention: the evaluation of a mass media campaign', Canadian Journal of Criminology 23, 191-202

Shaw, C.R. and McKay, H.D. (1929) Delinquency Areas, Chicago: University of Chicago Press

Shaw, C.R. and McKay, H.D. (1931) Social Factors in Juvenile Delinquency, Washington DC: US Government Printing Office

Shaw, C.R. and McKay, H.D. (1942) Juvenile Delinquency and Urban Areas, Chicago: University of Chicago Press

Shaw, C.R. and McKay, H.D. (1969) Juvenile Delinquency and Urban Areas, (rev. ed.), Chicago: University of Chicago Press

Shaw, C.R., and Moore, M.E. (1931) The Natural History of a Delinquent Career, Chicago: University of Chicago Press

Sherman, L.W. (1986) 'Policing communities: what works?', in A.J. Reiss and M. Tonry (eds), Communities and Crime, Chicago: University of Chicago Press, 343-86

Solicitor General of Canada (1984) 'Crime prevention: awareness and practice', Canadian Urban Victimization Survey Bulletin, 3, Ottawa, Ontario: Supply and Services Canada

Stoks, F. (1982) Assessing Urban Public Space Environments for Danger of Violent Crime - Especially Rape, PhD dissertation, Seattle, Washington: University of Washington

Taylor, R.B. and Gottfredson, S. (1986) 'Environmental design, crime, and prevention: an examination of community dynamics', in Reiss, A.J., Jr., and Tonry, M. (eds) Communities and Crime, Chicago: University of Chicago Press, 87-416

Taylor, R,. Gottfredson, S., and Brower, S. (1980) 'The defensibility of defensible space: a critical review', in T. Hirschi and M. Gottfredson (eds) Understanding Crime, Beverly Hills, California: Sage

Titus, R. (1982) Citizen and Environmental Crime Prevention, Washington, DC: National Institute of Justice

Titus, R. (1984) 'Residential burglary and the community response', in Clarke, R. and Hope, T. (eds), Coping With Burglary, Boston: Kluwer-Nijhoff, 97-130

Tobias, J. (1972) Nineteenth-Century Crime: Prevention and Punishment, Newton Abbot: David and Charles

Washnis, G.J. (1977) Citizen Involvement in Crime Prevention, Lexington: Lexington Books

Weaver, R., and Carroll, J. (1985) 'Crime perceptions in a natural setting by expert and novice shoplifters', Social Psychology Quarterly 48, 349-59

Wilson, J.Q. and Kelling, G.L. (1982) 'Broken windows: the

police and neighborhood safety', Atlantic, 249 (3), 29-38

White, T.W., Regan, K.J., Waller, J.D., and Wholey, J.S. (1975) Police Burglary Prevention Programs, Washington, DC: US Department of Justice

Worrell, P.B. (1984) An Evaluation of the Neighborhood Watch Program in Thunder Bay, Ottawa, Ontario: Ministry of the Solicitor General of Canada

Milton Keynes UK
Ingram Content Group UK Ltd.
UKHW031141141024
449569UK00024B/1169